ACCELERATED TESTING

A Practitioner's Guide to Accelerated and Reliability Testing

Other SAE titles of interest:

Automotive Engineering Fundamentals
By Richard Stone and Jeffrey K. Ball
(Order No. R-199)

Finite Element Analysis for Design Engineers
By Paul M. Kurowski
(Order No. R-349)

Finite Elements: Their Design and Performance
By Richard H. MacNeal
(Order No. B-608)

An Introduction to Modern Vehicle Design
By Julian Happian-Smith
(Order No. R-295)

The System Integration Process for Accelerated Development
By R.J. Menne and M.N. Rechs
(Order No. R-319)

Accelerated Testing

A Practitioner's Guide to Accelerated and Reliability Testing

Bryan Dodson

Harry Schwab

SAE INTERNATIONAL

For permission and licensing requests, contact:

SAE Permissions
400 Commonwealth Drive
Warrendale, PA 15096-0001 USA
E-mail: permissions@sae.org
Tel: 724-772-4028
Fax: 724-772-4891

Library of Congress Cataloging-in-Publication Data

Dodson, Bryan, 1962-
Accelerated testing : a practitioner's guide to accelerated and reliability testing / Bryan Dodson, Harry Schwab.
p. cm.
Includes bibliographical references and index.
ISB-10 0-7680-0690-2
ISBN-13 978-0-7680-0690-2
1. Reliability (Engineering). I. Schwab, Harry. II. Title.

TS173.D61 2006
620'.00452—dc22 2005057538

SAE International
400 Commonwealth Drive
Warrendale, PA 15096-0001 USA
E-mail: CustomerService@sae.org
Tel: 877-606-7323 (inside USA and Canada)
724-776-4970 (outside USA)
Fax: 724-776-1615

ISBN-13 978-0-7680-8042-1

SAE Order No. R-304

Printed in the United States of America.

ACKNOWLEDGMENTS

We are grateful to many individuals for helping prepare this book. Most notable are the reviewers. Several reviewers were anonymous, but you know who you are, and we thank you for your comments. Special thanks go to Lois Dodson and Matthew Dodson for creating the web site on the accompanying CD. We greatly appreciate Thermotron for providing the ESS material in Chapter 8, and the Quality Council of Indiana for allowing us to use some previously published material in Chapters 2 and 3.

PREFACE

This book presents accelerated testing from a practical viewpoint. The material is presented with the practitioner in mind; thus, proofs and derivations have been omitted, and numerous examples have been included. In addition, most examples have been worked in Microsoft® Excel and are included in the accompanying CD. For those desiring proofs and derivations, references are provided. Our goal is that practicing engineers will be able to apply the methods presented after studying this text.

Practitioners will find this text valuable as a comprehensive reference book, but this book is also ideal for use in college courses. In particular, it is recommended that this text be used for one-semester college courses. Students should have a familiarity with basic probability and statistics before attempting this material.

The text consists of eight chapters. Chapter 1 provides an introduction and overview of the limitations of accelerated testing. Chapters 2, 3, and 4 describe the fundamentals of statistical distributions, the most commonly used distributions in accelerated testing, and parameter estimation methods. Chapter 5 describes test plans for accelerated testing, including reliability growth. Chapter 6 explains models for accelerated aging, along with qualitative methods of accelerated testing. Chapter 7 explains environmental stress screening (ESS), and Chapter 8 presents the equipment and methods used in accelerated testing.

Be sure to use the accompanying CD, which contains a website to organize the material. The CD contains the following content:

- **Examples**—The examples presented in the text are worked in Microsoft Excel templates. These templates will be useful when applying the material to real-world problems.
- **Statistical Tables**—The statistical tables included in the appendices of books are holdovers to times when computers were not available. These tables give solutions to closed integrals of functions that could not be solved implicitly and required numerical methods to solve. These functions are now included in electronic spreadsheets. When the text references a value available in Appendix A, use the Microsoft Excel templates included on the CD.
- **Burn-In Optimization**—This is a Microsoft Excel template for determining the optimum burn-in duration based on the cost of burn-in time, burn-in failures, and field failures.
- **Random Number Generator**—This Microsoft Excel template generates random numbers that can be used for simulations. There are random number generators for the Weibull, normal, lognormal, and exponential distributions.

- **Government Documents—**This page contains many documents in PDF format. There are numerous military standards and handbooks related to reliability and accelerated testing.
- ***AMSAA Reliability Growth Handbook*—**This page provides the *AMSAA Reliability Growth Handbook* in Microsoft Word format.

Contents

Chapter 1

Introduction

The scientific theory of accelerated testing is highly developed, but the application of this theory has proven difficult, especially in the mobility industries. The required design life for many components exceeds 10 years, and the application environment is harsh and highly variable. Vehicles must operate reliably in arctic conditions and in desert conditions. Driving profiles range from the 16-year-old male to the 90-year-old female. An airliner may fly long-haul ocean routes for 20 years, while an identical model may fly short-range routes that result in many more takeoffs and landings over the life of the aircraft. Combining this variety into a realistic test that can be completed in a reasonable time frame with a reasonable budget is difficult and requires compromises.

The Purpose of Accelerated Testing

Ultimately, the only purpose of accelerated testing is cost reduction. The costs for accelerated testing—components, monitoring equipment, labor, test equipment, and so forth—must be recovered through lower warranty and the positive financial impact of customer satisfaction.

Accelerated tests fall into two categories: (1) development tests, and (2) quality assurance tests. During research, short inexpensive tests are needed to evaluate and improve performance. The progress of a product in these development tests is often monitored statistically with a reliability growth program. Some quality assurance tests are as follows:

- Design verification
- Production validation
- Periodic requalification

Quality assurance tests are often tied to statistical sampling plans with requirements such as a demonstrated reliability of at least 95% at 10 years in service with a confidence level of 90%. Statistically, 95% reliability with 90% confidence can be demonstrated by testing 45 units to the equivalent of 10 years in service. Table 1.1 gives the required sample sizes for some common reliability requirements.

Before proceeding with a test of 299, 45, or even 16 samples, the purpose of the test should be investigated. What does it cost to test 299 units? The following costs should be considered:

- Prototype costs
- Instrumentation costs (results monitoring)

TABLE 1.1
RELIABILITY DEMONSTRATION SAMPLE SIZES

Reliability	Confidence	Sample Size
99%	95%	299
99%	90%	229
99%	50%	69
95%	95%	59
95%	90%	45
95%	80%	31
90%	90%	22
90%	80%	16

- Setup costs
- Labor costs
- Laboratory costs (many tests take two or more months to complete)

Using the sample sizes shown in Table 1.1 allows no failures. If a failure occurs, do timing and budget constraints allow changes and a repeat of the test? What are the implications of bringing a product to market if that product did not demonstrate the required reliability with an appropriate level of confidence?

Design Life

Determining the design life to be simulated with an accelerated test can be difficult. Many automobile manufacturers specify a design life of 10 years for brake systems (excluding pads), but how does 10 years in service translate to an accelerated test? According to a study of brake system usage for minivans in 1990, the following statements are true:

- The median number of brake applies was 322,000 for 10 years in service.
- Five percent of the vehicles had more than 592,000 brake applies in 10 years of service.
- One percent of the vehicles had more than 709,000 brake applies in 10 years of service.
- The force of the brake apply ranged from 0.1 to 1.0 g-force, with the shape of the distribution shown in Figure 1.1.

How many times should the brake system be cycled in the accelerated test representing 10 years? Engineers design for the most stressful conditions; therefore, does this mean that the number of cycles is determined by the most stressful driver?

User profiles are often defined by percentile. The 95th percentile point is the point with 5% of the users having a more stressful profile. One percent of the users have a driving profile that is

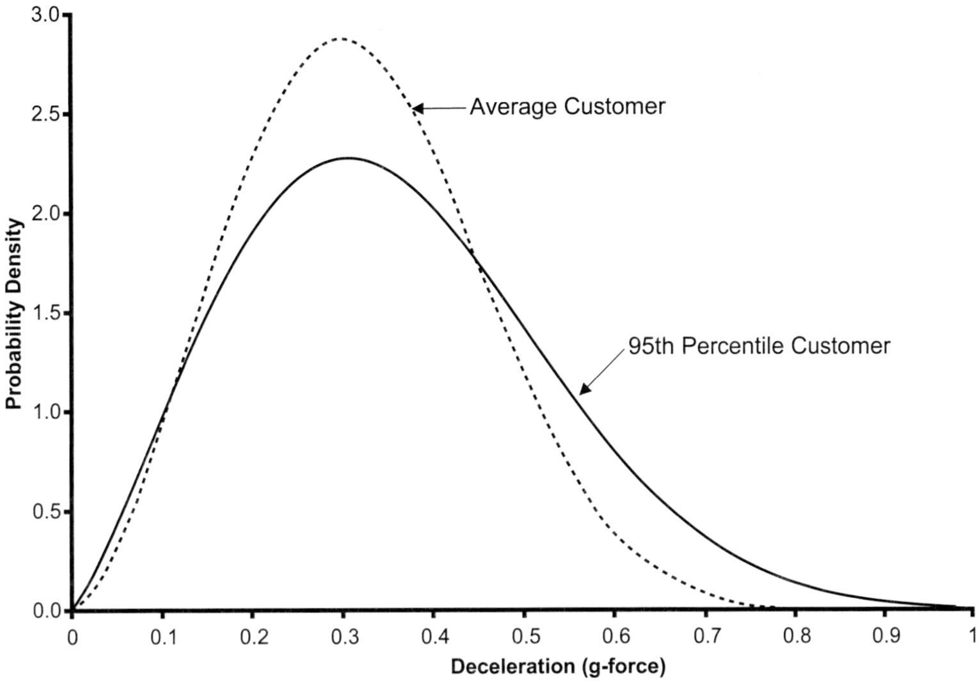

Figure 1.1 *Distribution of brake force application.*

more stressful than the 99th percentile driver. Table 1.2 gives the number of brake applications as a function of the user percentile; these data also are shown in Figure 1.2.

TABLE 1.2
PERCENTILES FOR MINIVAN BRAKE APPLICATIONS

Percentile	Number of Brake Applications
50th	321,891
60th	361,586
70th	405,155
75th	429,673
80th	457,241
85th	489,671
90th	530,829
95th	592,344
97.5th	646,007
99th	708,571
99.9th	838,987

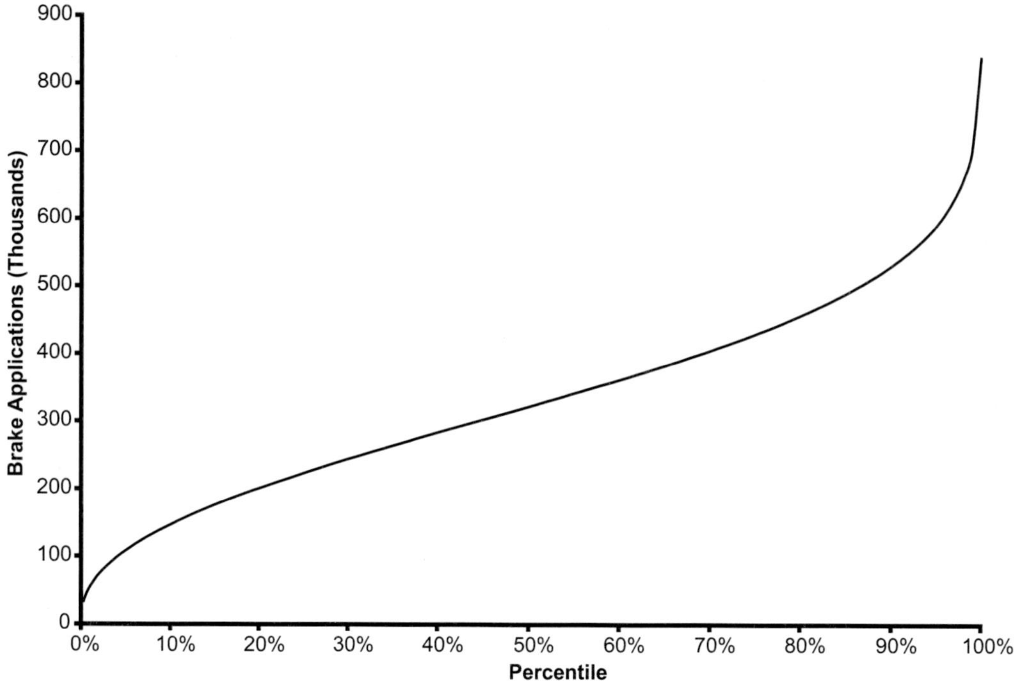

Figure 1.2 *Percentiles for minivan brake applications.*

As shown in Figure 1.2, the number of brake applications increases dramatically as the percent of the population covered nears 100%. This is typical of many other applications, such as door slams, ignition cycles, and trunk release cycles. To increase the percent of the population covered from 75% to 99.9 % requires an approximate doubling of the number of cycles in the accelerated test. Not only does this increase the cost and duration of the test, but the cost of the component increases because the number of cycles in the test is part of the design requirement.

The percent of the population covered is a compromise among development cost, development time, component cost, and the field performance of the component. For safety-critical items, the user percentile may exceed 100% to allow a safety margin. For other items, such as glove box latches, the user percentile may be as low as 80%. In reality, there is no 95th percentile user. There is a 95th percentile user for number of cycles, a 95th percentile user for temperature, a 95th percentile user for number of salt exposure, a 95th percentile user for vibration, and so forth. However, determining the 95th percentile user for the combination of conditions is unrealistic.

The worst-case user profile may not be at the high end for the number of cycles of operation. Consider a parking brake. The worst case may be a brake that is used for the first time after the vehicle is 10 years old. This type of user profile must be incorporated into a test separate from a test utilizing the 95th percentile of parking brake applications.

Accelerating a test by eliminating the time between cycles can introduce unrealistic conditions. Consider a durability test for an automobile door. The door is opened and closed 38,000 times in 12 hours. Opening and closing the door this quickly does not allow the door hinges or latches to cool, nor does it give any contaminants that may be introduced in the hinges time to form corrosion. Consider an automobile engine: the 95th percentile user profile for engine on-time is approximately 7,000 hours. Does running the engine for 7,000 consecutive hours approximate 7,000 hours of operation over 10 years? Consider an automobile starter: the 95th percentile user profile for the number of engine starts is approximately 4,000. Starting the engine 4,000 times as quickly as possible does not stress the starter as much as actual usage conditions because the engine would be warm for nearly every engine start. To more adequately represent true usage conditions, the engine would need to be cooled for some of the starts.

Statistical Sample Size Determination

The sample sizes given in Table 1.1 are based on statistical sampling. Statistical confidence assumes a random sample representative of the population. Obtaining a random sample representative of the population requires all sources of variation to be present, such as the following:

- Variation from multiple production operators
- Variation from multiple lots of raw materials
- Variation from tool wear
- Variation from machine maintenance
- Variation from seasonal climatic changes
- Variation from supplier changes

It may be possible to obtain a random sample representative of the population for periodic requalifications, but it is nearly impossible for new product development. Thus, designing tests to demonstrate reliability with statistical confidence is not always possible. The best alternative is to test with worst-case tolerances.

Tolerances in Accelerated Testing

Determining the worst-case combination of tolerances can be difficult. Consider the simple system shown in Figure 1.3. Component A is inserted into Component B and rotates during operation. The worst-case tolerance is either Component A at maximum diameter and the inner diameter of Component B at a minimum, or Component A at a minimum diameter and the inner diameter of Component B at a maximum.

But even with this simple system, other tolerances must be accounted for, such as the following:

- Surface finish (for both components)
- Volume of lubricant
- Viscosity of lubricant
- Roundness (for both components)
- Hardness (for both components)

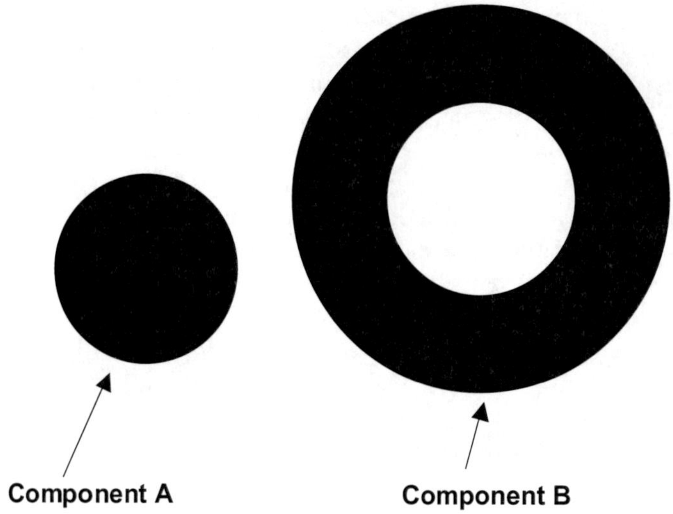

Figure 1.3 *Simple tolerancing example.*

The number of tolerance combinations can become unmanageable. Table 1.3 shows the number of possible tolerance combinations as a function of the number of dimensions. With 10 characteristics to consider for worst-case tolerancing in this simple two-component system, there are more than 1,000 combinations of tolerances to consider. Determining which of these 1,000 combinations is the worst case is often difficult.

TABLE 1.3
NUMBER OF TOLERANCE COMBINATIONS

Number of Characteristics	Number of Tolerance Combinations
2	4
3	8
4	16
5	32
10	1,024
20	1,048,576
50	$1.126 (10^{15})$
100	$1.268 (10^{30})$

Confounding the problem is the fact that the worst-case tolerance combination for a specific environmental condition may be the best-case tolerance combination for another environmental condition. Manufacturing capabilities also complicate testing at worst-case tolerance combinations. It is often impossible or prohibitively expensive to produce parts at the desired tolerance level. In these cases, a compromise is made by using a dimension as close as possible to the desired value.

Ideally, if all characteristics are within tolerance, the system would work perfectly and survive for the designed life. And if one or more characteristics are out of tolerance, the system would fail. Reality demonstrates that a component with a characteristic slightly out of tolerance is nearly identical to a component with the same characteristic slightly within tolerance. Tolerances are not always scientifically determined because time and budget do not always allow for enough research. There is a strong correlation between the defect rate in the manufacturing facility and field reliability. A portion of the reduction in defect rate has been due to a reduction of manufacturing variability. As manufacturing variability is reduced, characteristics are grouped closer to the target.

Consider a motor with its long-term durability dependent on the precision fit of three components in a housing. The three components are stacked in the housing; historically, the tolerance stackup has caused durability problems, and the maximum stackup of the three components has been specified at 110. To meet this requirement, an engineer created the specifications shown in Table 1.4.

TABLE 1.4
MOTOR COMPONENT TOLERANCES

Component	A	B	C	Total
Target Size	30	20	10	60
Maximum Size	50	30	15	95

If the three components are manufactured to the target, the total stackup is 60. However, there is always variance in processes, so the engineer specifies a maximum allowable size. If the manufacturing capability for each of the components is 3 sigma (a defect rate of 67,000 parts per million), the process will produce the results shown in Figure 1.4 for the stackup of the system.

By increasing the manufacturing capability for each of the components to 4 sigma (a defect rate of 6,200 parts per million), the process will produce the results shown in Figure 1.5 for the stackup of the system.

The motor housing has a perfect fit with the three components if the stackup is 60. Any deviation from 60 will reduce the life of the motor. As long as the total stackup is less than 110, the motor will have an acceptable life; however, motors with a stackup closer to 60 will last longer. It is easy to see that the reduced variance in manufacturing will increase the life of the motors.

Manufacturing capability cannot be overlooked by reliability engineers. First-time capability verification, statistical process control (SPC), and control plans are essential to providing highly reliable products. Without capable manufacturing, all previous reliability efforts will provide little or no benefit.

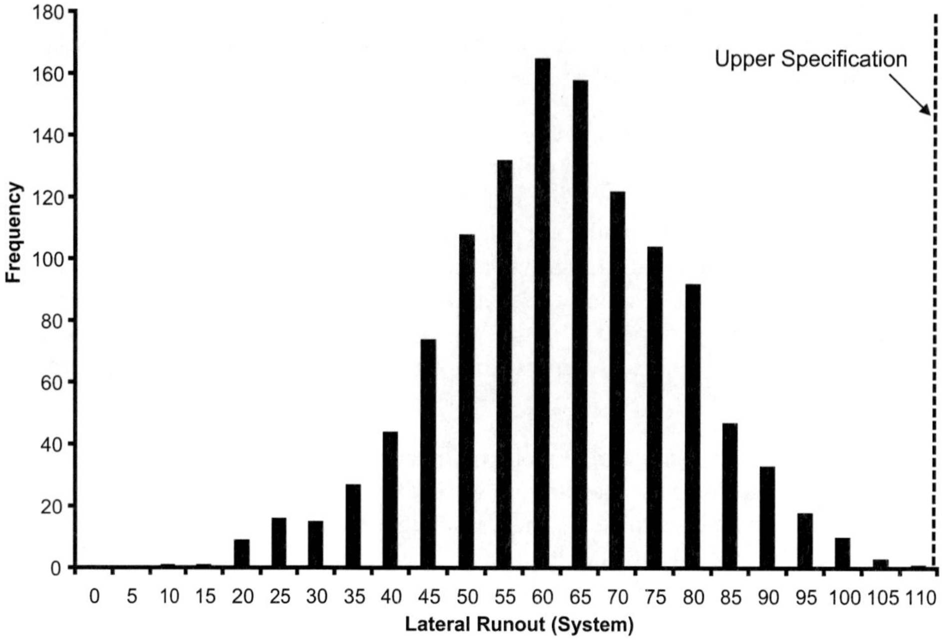

Figure 1.4 *Tolerance stackup at a 3-sigma quality level.*

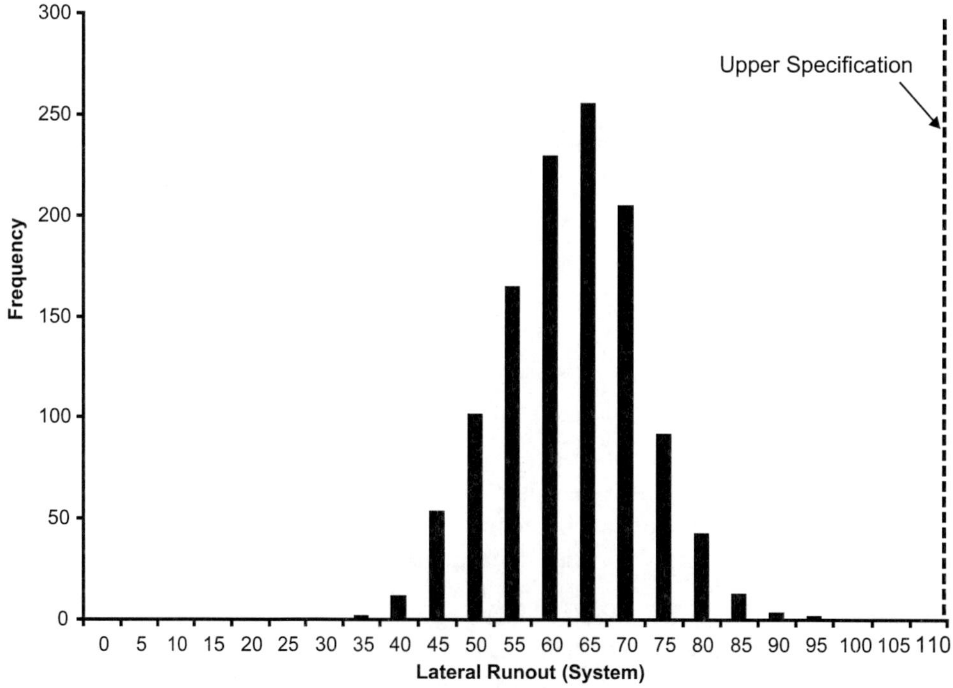

Figure 1.5 *Tolerance stackup at a 4-sigma quality level.*

Financial Considerations

Priorities for a reliability program are determined the same way as most other programs. The number one priority is an emergency. If there is a hole in one of the water pipes in your home, you will find a way to fix it, regardless of your financial situation.

The next level of priority is an obvious need that can be delayed with a risk. Consider again the leaking water pipe. If the water pipe is patched with duct tape and epoxy, and while fixing the pipe it is determined that all the water pipes in the home are in danger of bursting, then obviously there is a need to replace all the water pipes in the house. However, this is an expensive task and can be delayed. There is no immediate crisis, but by delaying the repair, there is a risk of an expensive accident. If a water pipe bursts, thousands of dollars of damage will result. This risk is tolerated because the immediate expense of correcting the problem is perceived to be greater than the cost of the water pipes bursting weighted by the probability of the water pipes bursting.

The most dangerous priority is one that is not known. Consider a home that is being consumed by termites without the owner's knowledge. Nothing is done to correct the problem because the owner is ignorant of the problem. For reliability programs, the largest expenses are often overlooked.

A typical reliability program contains the following eight elements:

1. **Understand your customer's requirements.**

   a. Environment.

      i. Vibration (g-force root mean square [GRMS] and frequency profile).

      ii. Humidity.

      iii. Temperature (absolutes and cycling rates).

      iv. Electrical stimulus (electrostatic discharge [ESD]).

      v. Contamination (salt, brake fluid, transmission fluid, milkshakes).

   b. Duty cycle.

      i. Number of usage cycles.

      ii. Required life in months or years.

   c. Load.

      i. Pounds of force.

      ii. Pressure.

      iii. Voltage.

   iv. Current.

   d. Reliability goals.

2. **Feedback from similar components.**

   a. FRACAS (failure rate analysis and corrective action system)—Parts from test failures, internal production failures, external production failures, and field returns must be analyzed and cataloged.

   b. J.D. Power and Associates.

   c. Warranty (return rates, feedback from customers and technicians).

   d. Development tests, design verification tests, and production validation tests.

3. **Begin the FMEA (failure modes and effects analysis) process. The FMEA will be updated during the entire process.**

4. **Intelligent design.**

   a. Use design guides—All lessons from previous incidents must be captured in design guides. This includes all information from the FRACAS.

   b. Parameter design—Choose the design variable levels to minimize the effect of uncontrollable variables.

   c. Tolerance design—Scientifically determine the correct drawing specifications.

   d. Hold periodic design reviews.

   e. Design with information from developmental activities.

      i. Sneak circuit analysis.

      ii. Highly accelerated life testing (HALT).

      iii. Step-stress tests to failure.

      iv. Worst-case tolerance analysis.

5. **Concept validation (includes parts supplied by outside sources).**

   a. Early in the development phase, have short, inexpensive tests to provide approximate results. The purpose of these tests is to provide engineering feedback.

   b. Every concept must pass an independent (i.e., not conducted by engineering) verification test. The concept should include design limits. For example, the component has been validated to operate up to 85°C, withstand brake fluid, and 4.8 GRMS random vibration in a frequency range from 0 to 800 Hz.

c. A standard list of validated subsystems and components should be maintained. This includes parts supplied from outside sources.

6. Design validation (includes parts supplied by outside sources).

a. Early in the development phase, have short, inexpensive tests to provide approximate results. The purpose of these tests is to provide engineering feedback.

b. Every design must pass an independent (not conducted by engineering) verification test. Be careful not to burden the company with timing and cost issues when specifying the test. Build on the results of the concept verification and any other implementations of the concept.

c. System simulation.

7. Manufacturing.

a. Parts from production intent tooling must pass the design validation test.

b. The production process is qualified by producing a specified number of parts at production rates, and obtaining a minimum C_{pk} of 1.67 for every drawing specification.

c. Ensure compliance with an SPC program.

i. All drawing specifications will be monitored with an electronic SPC system.

ii. For the first week of production, the sampling rate is 100%.

iii. If a C_{pk} of 1.67 is achieved for the first week, the sampling rate may be reduced.

iv. Each drawing specification must have a control plan that details the critical processes affecting the drawing specification. Each of these processes also must be monitored with an SPC.

v. For each measurement system:

1. Provide a control plan to ensure stability and bias control.

2. The measurement error should be an acceptable percentage of the tolerance. What is acceptable is a function of the process C_{pk}.

d. Qualify the storage, transportation, and installation systems.

i. Can parts be contaminated during storage?

ii. Can parts degrade during storage?

1. Rubber ages.

2. Gas charge levels decrease.

iii. Are the temperature and vibration profiles during transportation significantly different from those of the vehicle specification?

iv. Is the part protected from corrosion caused by the salt in the air during transportation on an ocean liner?

v. Requalify if the transportation system is changed.

8. Change control—Any changes in engineering, production, or the supply base must be qualified.

The following is a consolidated set of responses based on actual experiences.

Response from Finance:

After analyzing a reliability program, there appear to be many opportunities for savings.

1. The current understanding of the customer's environment is adequate. There is no need to collect additional customer usage or environmental data.
2. How many prototypes will be used during the parameter design and tolerance design processes? What is the cost of a prototype?
3. What is the cost of testing?
   a. HALT
   b. Design verification
   c. Production validation
   d. Step-stress testing
4. The manufacturing qualification plan appears to be excessive. We recommend reducing the number of samples required for initial validation and reducing the number of characteristics that require control plans.

Response from Engineering:

Although we agree that the approach of Reliability is correct, we are supporting Finance. The program is behind schedule, and reducing the reliability effort will improve program timing. In addition, with recent cutbacks (or the recent increase in business), Engineering lacks the resources to complete the entire reliability program.

Response from Manufacturing:

The program proposed by Reliability will ensure initial quality, but it is too costly. With recent [enter any excuse here], manufacturing cannot meet the initial production schedule with the restrictions proposed by Reliability. The reliability program would also require additional personnel to maintain the SPC program.

These responses result in an organization with a strong incentive to gamble that there will be no consequences for abandoning a thorough reliability program in favor of a program that is less expensive. This behavior sub-optimizes the finances of the company by assuming any potential failure costs are near zero. To be effective, a good reliability program must include a financial assessment of the risks involved if reliability activities are not completely executed.

The urge to reduce the investment in the reliability program can be combatted by visualizing the failure costs that the reliability program is designed to prevent. In addition to warranty costs, other failure costs are as follows:

- Customer returns
- Customer stop shipments
- Retrofits
- Recalls

These costs often are ignored because they are not quantified. An effective method for quantifying these costs is to record a score for each element in the reliability program and compare this to the field performance of the product. This can be done by auditors using a grade scale of A through F for each element of the program. The grades for each element can be combined into a grade point average (GPA) for the program using 4 points for an A, 3 points for a B, and so forth.

Table 1.5 gives an example of how a reliability program may be scored, and Table 1.6 shows how the field performance is recorded. When quantifying these costs, be sure to include all associated labor and travel costs. For example, a customer may charge $500 for returning a single part; however, the associated paperwork, travel, and investigation could easily be several thousand dollars.

TABLE 1.5
EXAMPLE RELIABILITY PROGRAM SCORES

Reliability Program Item	Score (GPA)
Understanding of Customer Requirements	B–3
FMEA	A–4
FRACAS	C–2
Verification	C–2
Validation	D–1
Manufacturing	B–3
Overall Program Score	**2.33**

TABLE 1.6
EXAMPLE FIELD RELIABILITY PERFORMANCE

Reliability Performance Item	Cost
Customer Returns	$8,245
Customer Stop Shipments	$0
Retrofits	$761,291
Recalls	$0
Overall Program Unreliability Cost	**$769,536**

Figure 1.6 is a scatter chart of the results of several programs. The slope of the trend line quantifies the loss when the reliability program is not fully executed. For this example, moving the GPA of the overall reliability program by one point is expected to result in a savings of $755,000 in failure costs. This savings can be used to financially justify the investment in the reliability program.

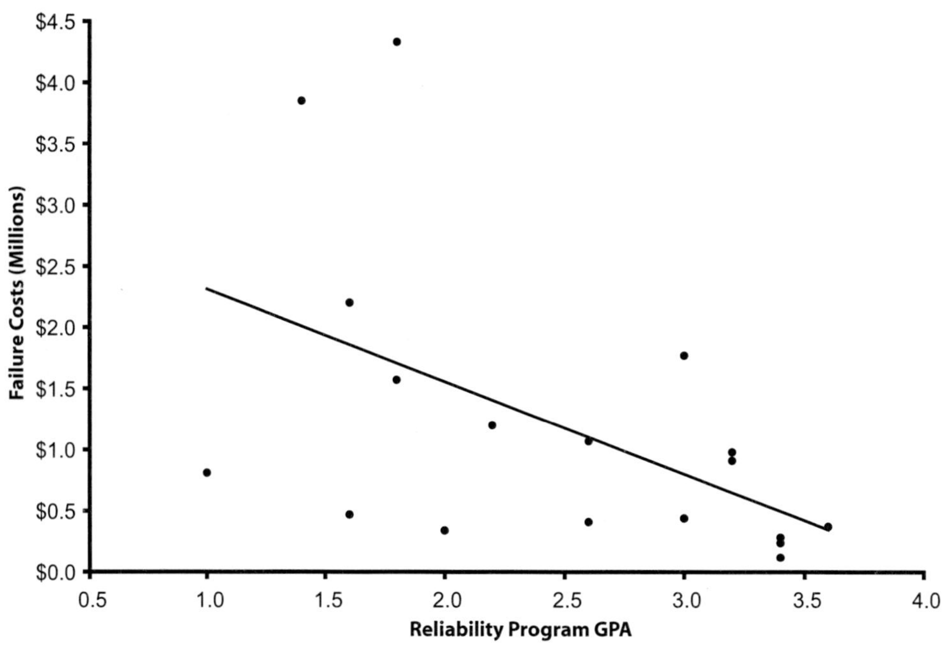

Figure 1.6 *Reliability program execution score versus failure costs.*

These failure costs are similar to the cost of water pipes bursting in your house. You know of the risk, and you decide to act on the risk or tolerate the risk, based on the finances of the situation.

Another method to focus management's attention on reliability is by presenting the effects of the data shown in Table 1.6 on corporate profits. The data in Figure 1.7 are examples of the effects of a poor reliability program. Money was saved years earlier by gambling with a substandard reliability program, but as shown in Figure 1.7, the short-term gain was not a good long-term investment.

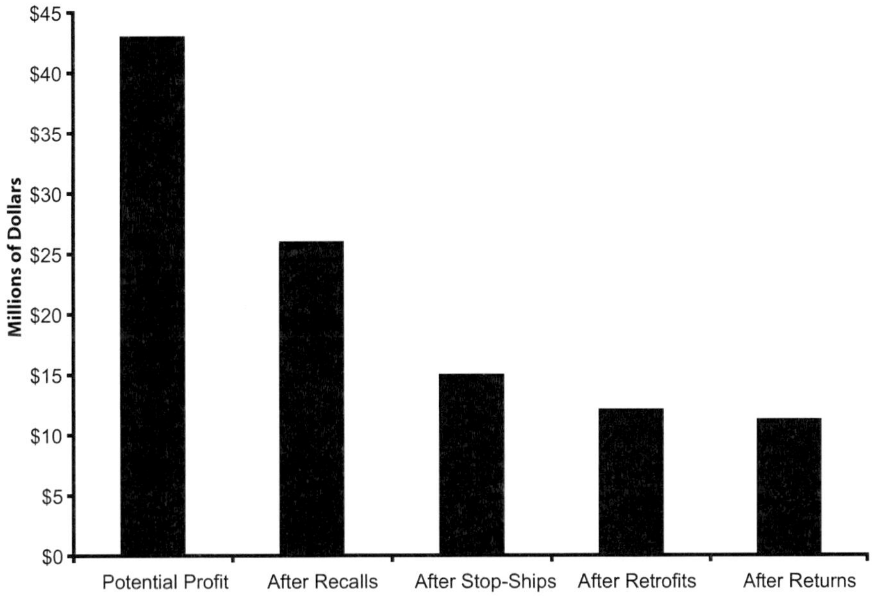

Figure 1.7 *Effect of poor reliability on company profits.*

Similar to termites damaging a home without the owner's knowledge, hidden reliability costs are causing poor decisions to be made and are damaging profits. The losses caused by these hidden costs can be orders of magnitude greater than warranty costs. To illustrate this concept, consider the automotive industry.

For model year 1998, the average vehicle manufactured by General Motors, Ford, or Chrysler (the "Big Three") required $462* in repairs. These automakers sell approximately 13 million vehicles in North America annually, resulting in a total warranty bill of $6 billion. That may sound like a lot of money, but it is by far the smallest piece of the cost of poor reliability. Table 1.7 illustrates the retail value for several 1998 model year vehicles with sticker prices all within a $500 range.

For lease vehicles, the manufacturer absorbs the $5,715 difference in resale value between Vehicle B and Vehicle H. For non-lease vehicles, the owner of Vehicle B absorbs the cost. But

* Day, Joseph C., address to the Economic Club of Detroit, December 11, 2000.

TABLE 1.7
VEHICLE RESALE VALUE

Vehicle (1998 Model Year)	Retail Value as of July 2001	*Consumer Reports* Reliability Rating*
A	$8,430	–45
B	$9,500	20
C	$9,725	18
D	$11,150	25
E	$11,150	30
F	$13,315	–5
G	$14,365	55
H	$15,215	50

* The *Consumer Reports* scale is from –80 to 80, with –80 being the worst and 80 being the best.

this does not mean the manufacturer is not impacted. The reduced retail value is reflected in the ability of the manufacturer to price new vehicles. The manufacturer of Vehicle H can charge more for new vehicles because they depreciate more slowly. Considering that sales for many of these midsized sedans topped 200,000 units, the $5,715 difference in resale value is worth more than $1 billion annually. Figure 1.8 shows the correlation of the reliability of a vehicle and its resale value. Using the slope of the regression line shown in Figure 1.8, a single point in *Consumer Reports*' reliability rating is worth $51.58.

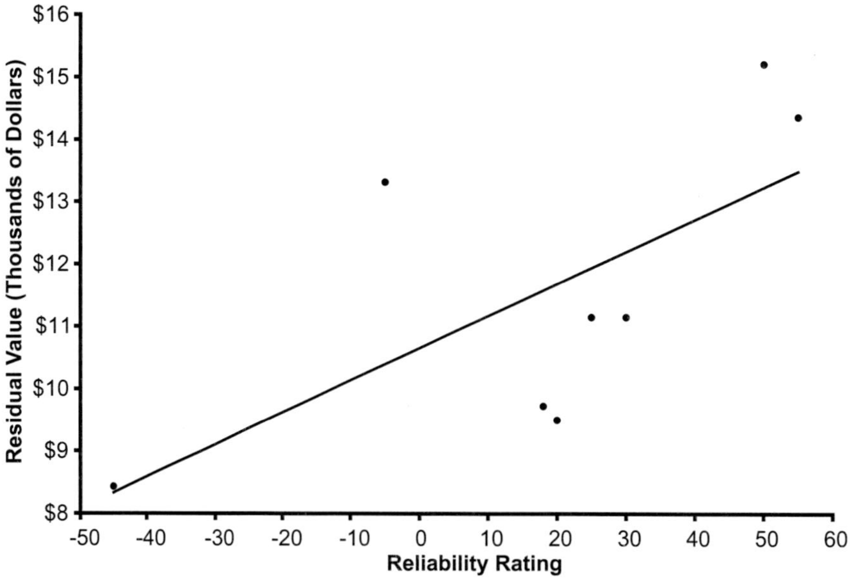

Figure 1.8 *Automobile resale value versus reliability rating.*

When faced with pressures to compromise on your reliability program, consider the system cost. For example, Design A uses a material that is a nickel less expensive than Design B, and the component is expected to be installed on 300,000 vehicles. If Design A is chosen, the savings is \$15,000 in material cost. This is a good decision only if the expected impact to warranty, resale value, and market share is less than \$15,000. Before using Design A, verify the reliability of the design by doing the following:

- Compare the designs using a durability test.
- Ensure your durability test will incorporate any potential failure modes associated with material changes.
- Verify key performance issues with computer modeling.
- Check for any potential system failures that could be caused by a material change.
- Explore the impact on manufacturing. What is the impact of material changes, processing changes, and so forth? Are there issues such as dunnage issues, environmental issues, and safety hazards?

Summary

The theoretical science of accelerated testing is exact, but implementing an accelerated testing plan requires several compromises. Ultimately, the science involved is inexact and serves only as a guideline for engineering judgment. Reliability demonstration based on statistically determined sample sizes is often invalid because the samples could not be a random representation of production parts. Testing to worst-case tolerance limits is difficult because of the number of combinations and the difficulty of producing parts at the desired tolerance level.

Automobile manufacturers often specify a design life of 10 years for automobiles. Many types of aircraft have a design life of 25 years, and some B-52 bombers have been in service for more than 40 years. Accelerating a test for an automobile component by a factor of 10 would yield a test with a one-year duration. This obviously is unacceptable. Obtaining a test with a duration of one month requires an acceleration factor of 120. A test lasting 24 hours would have an acceleration factor of 3,653.

Is this type of acceleration possible? Although this is a slightly controversial subject, many experts suggest it is impossible to accelerate a test by more than a factor of 10 without losing some correlation to real-world conditions. This is another of the ambiguities faced when accelerating testing conditions. A test is required, and the test is useless if it cannot be completed in a reasonable time frame. However, the greater the acceleration, the less realistic the test.

Accelerated testing is a balance between science and judgment. Do not let the science cause bad decisions to be made. For example, if demonstrating 95% reliability at 150,000 miles in service calls for testing 45 units for 2,000 hours without failure, do not be concerned if only 35 units can be tested. The sample size of 45 assumes random sampling from a population representative of production. Violating this assumption is more important than testing with a reduced sample size. In this situation, try to ensure that the test is representative of real-world conditions, and secure 35 samples with key characteristics set at worst-case tolerance levels.

Chapter 2

Probability Fundamentals

Statistical methods are based on random sampling. Four functions are used to characterize the behavior of random variables:

1. The probability density function
2. The cumulative distribution function
3. The reliability function
4. The hazard function

If any one of these four functions is known, the others can be derived. This chapter describes these functions in detail.

Sampling

Statistical methods are used to describe populations by using samples. If the population of interest is small enough, statistical methods are not required; every member of the population can be measured. If every member of the population has been measured, a confidence interval for the mean is not necessary, because the mean is known with certainty (ignoring measurement error).

For statistical methods to be valid, all samples must be chosen randomly. Suppose the time to fail for 60-watt light bulbs is required to be greater than 100 hours, and the customer verifies this on a monthly basis by testing 20 light bulbs. If the first 20 light bulbs produced each month are placed on the test, any inferences about the reliability of the light bulbs produced during the entire month would be invalid because the 20 light bulbs tested do not represent the entire month of production. For a sample to be random, every member of the population must have an equal chance of being selected for the test.

Now suppose that when 20 light bulbs, randomly selected from an entire month's production, are placed on a test stand, the test is ended after 10 of the light bulbs fail. Table 2.1 gives an example of what the test data might look like.

What is the average time to fail for the light bulbs? It obviously is not the average time to fail of the 10 failed light bulbs, because if testing were continued until all 20 light bulbs failed, 10 data points would be added to the data set that are all greater than any of the 10 previous data points. When testing is ended before all items fail, the randomness of the sample has been destroyed. Because only 10 of the light bulbs failed, these 10 items are *not* a random sample that is representative of the population. The initial sample of 20 items is a random sample

TABLE 2.1
TIME TO FAIL FOR LIGHT BULBS

23	63	90
39	72	96
41	79	
58	83	
10 units survive for 96 hours without failing.		

representative of the population. By ending testing after 10 items have failed, the randomness has been destroyed by systematically selecting the 10 items with the smallest time to fail.

This type of situation is called censoring. Statistical inferences can be made using censored data, but special techniques are required. The situation described above is right censoring; the time to fail for a portion of the data is not known, but it is known that the time to fail is greater than a given value. Right-censored data may be either time censored or failure censored. If testing is ended after a predetermined amount of time, it is time censored. If testing is ended after a predetermined number of failures, the data are failure censored. The data in Table 2.1 are failure censored. Time censoring is also known as Type I censoring, and failure censoring is also known as Type II censoring.

The opposite of right censoring is left censoring. For a portion of the data, the absolute value is not known, but it is known that the absolute value is less than a given value. An example of this is an instrument used to measure the concentration of chemicals in solution. If the instrument cannot detect the chemical below certain concentrations, it does not mean there is no chemical present, but that the level of chemical is below the detectable level of the instrument.

Another type of right censoring is multiple censoring. Multiple censoring occurs when items are removed from testing at more than one point in time. Field data are often multiple censored. Table 2.2 gives an example of multiple censoring. A "+" next to a value indicates the item was removed from testing at that time without failing.

TABLE 2.2
MULTIPLE CENSORED DATA

112	172	220
145 +	183	225 +
151	184 +	225 +
160	191	225 +
166 +	199	225 +

There are two types of data: (1) continuous, and (2) discrete. Continuous variables are unlimited in their degree of precision. For example, the length of a rod may be 5, 5.01, or 5.001 inches.

It is impossible to state that a rod is exactly 5 inches long—only that the length of the rod falls within a specific interval. Discrete variables are limited to specific values. For example, if a die is rolled, the result is either 1, 2, 3, 4, 5, or 6. There is no possibility of obtaining any value other than these six values.

Probability Density Function

The probability density function, $f(x)$, describes the behavior of a random variable. Typically, the probability density function is viewed as the "shape" of the distribution. Consider the histogram of the length of fish as shown in Figure 2.1.

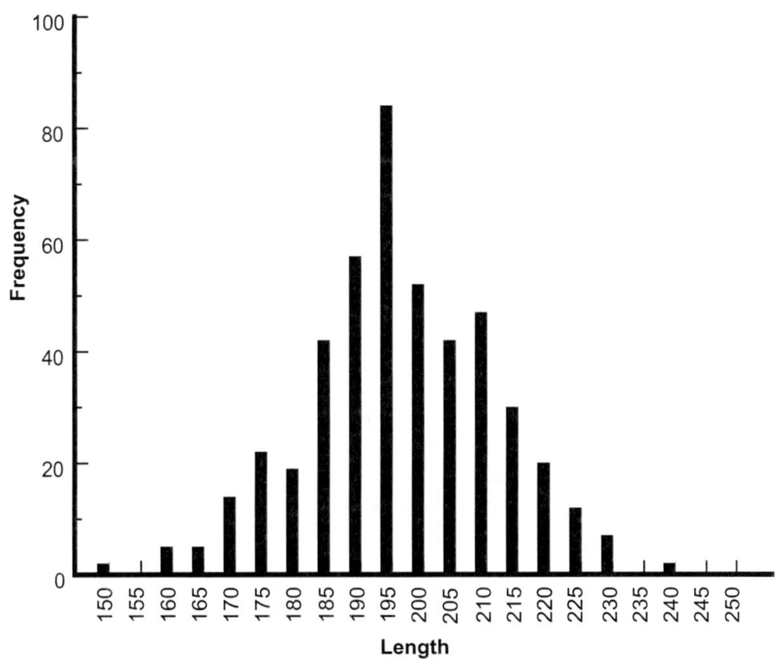

Figure 2.1 *Example histogram.*

A histogram is an approximation of the shape of the distribution. The histogram shown in Figure 2.1 appears symmetrical. Figure 2.2 shows this histogram with a smooth curve overlying the data. The smooth curve is the statistical model that describes the population—in this case, the normal distribution. When using statistics, the smooth curve represents the population, and the differences between the sample data represented by the histogram and the population data represented by the smooth curve are assumed to be due to sampling error. In reality, the differences could also be caused by lack of randomness in the sample or an incorrect model.

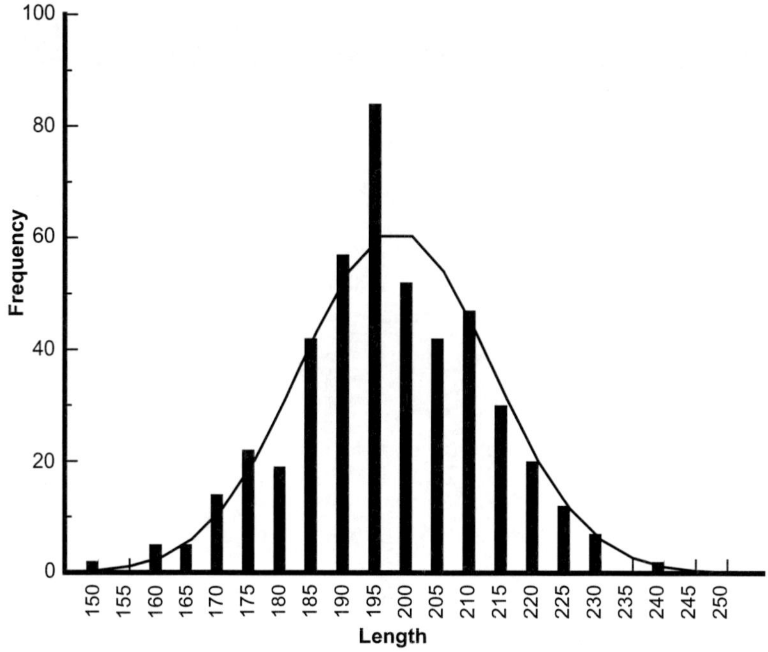

Figure 2.2 *Example histogram with overlaid model.*

The probability density function is similar to the overlaid model shown in Figure 2.2. The area below the probability density function to the left of a given value x is equal to the probability of the random variable represented on the x-axis being less than the given value x. Because the probability density function represents the entire sample space, the area under the probability density function must equal 1. Negative probabilities are impossible; therefore, the probability density function, $f(x)$, must be positive for all values of x. Stating these two requirements mathematically,

$$\int_{-\infty}^{\infty} f(x) = 1 \tag{2.1}$$

and $f(x) \geq 0$ for continuous distributions. For discrete distributions,

$$\sum^{n} f(x) = 1 \tag{2.2}$$

for all values of n, and $f(x) \geq 0$.

The area below the curve in Figure 2.2 is greater than 1; thus, this curve is not a valid probability density function. The density function representing the data in Figure 2.2 is shown in Figure 2.3.

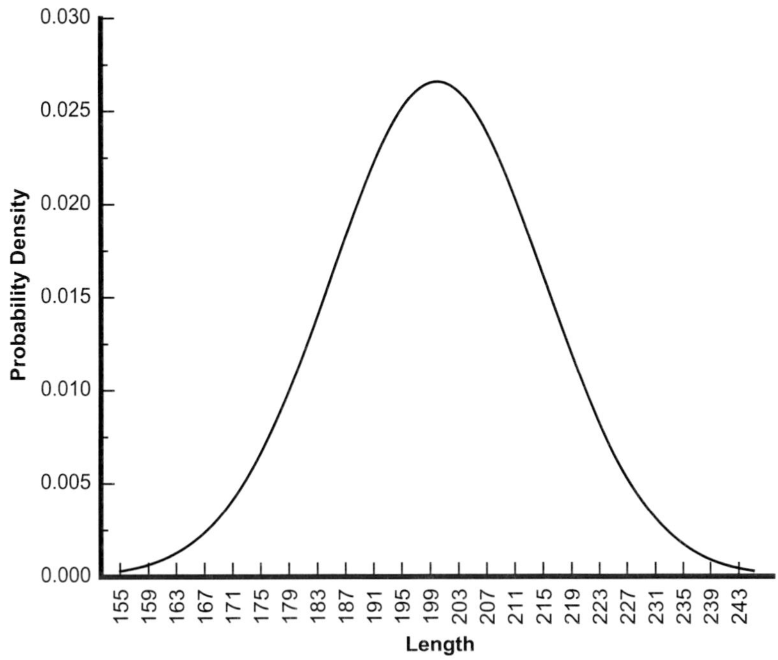

Figure 2.3 *Probability density function for the length of fish.*

Figure 2.4 demonstrates how the probability density function is used to compute probabilities. The area of the shaded region represents the probability of a single fish, drawn randomly from the population, having a length less than 185. This probability is 15.9%.

Figure 2.5 demonstrates the probability of the length of one randomly selected fish having a length greater than 220.

The area of the shaded region in Figure 2.6 demonstrates the probability of the length of one randomly selected fish having a length greater than 205 and less than 215. Note that as the width of the interval decreases, the area, and thus the probability of the length falling in the interval, decreases. This also implies that the probability of the length of one randomly selected fish having a length exactly equal to a specific value is 0. This is because the area of a line is 0.

Example 2.1: A probability density function is defined as

$$f(x) = \frac{a}{x}, \quad 1 \le x \le 10$$

For $f(x)$ to be a valid probability density function, what is the value of a?

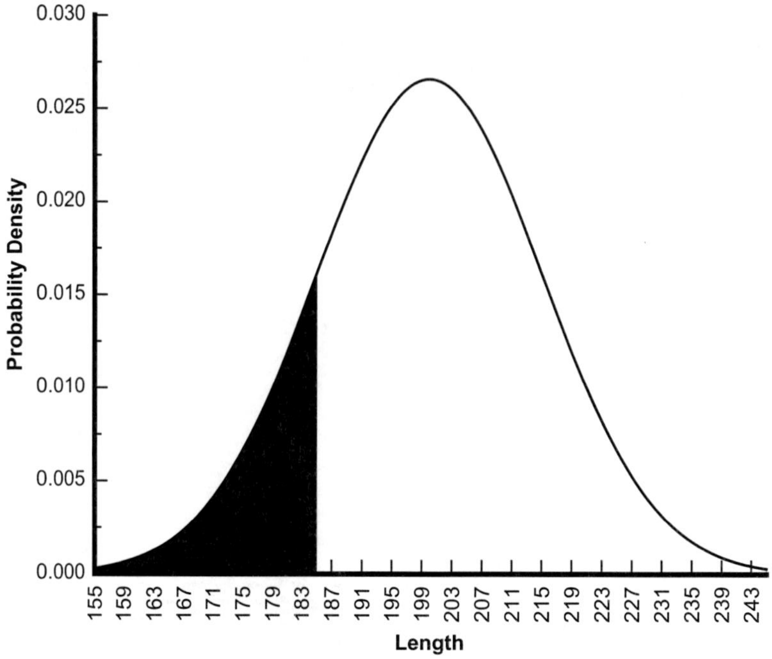

Figure 2.4 *The probability of the length being less than 185.*

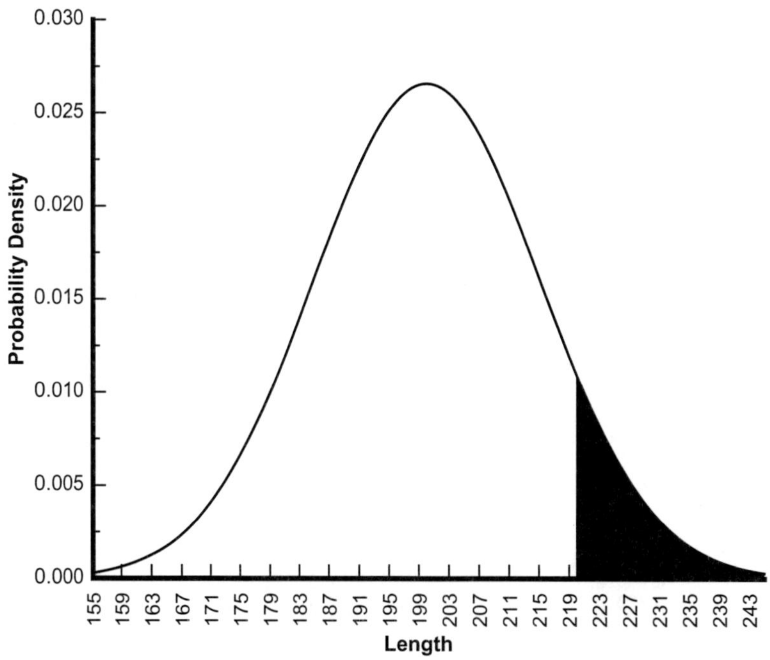

Figure 2.5 *The probability of the length being greater than 220.*

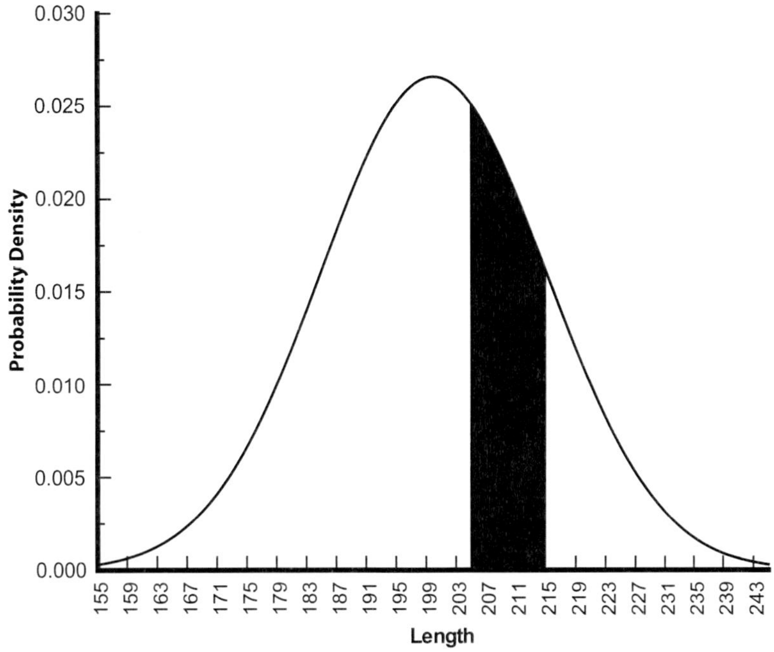

Figure 2.6 *The probability of the length being between 205 and 215.*

Solution: To be a valid probability density function, all values of $f(x)$ must be positive, and the area beneath $f(x)$ must equal 1. The first condition is met by restricting a and x to positive numbers. To meet the second condition, the integral of $f(x)$ from 1 to 10 must equal 1.

$$\int_1^{10} \frac{a}{x} = 1$$

$$a \ln x\Big|_1^{10} = 1$$

$$a \ln(10) - a \ln(1) = 1$$

$$a = \frac{1}{\ln(10)}$$

Cumulative Distribution Function

The cumulative distribution function, $F(x)$, denotes the area beneath the probability density function to the left of x, as shown in Figure 2.7.

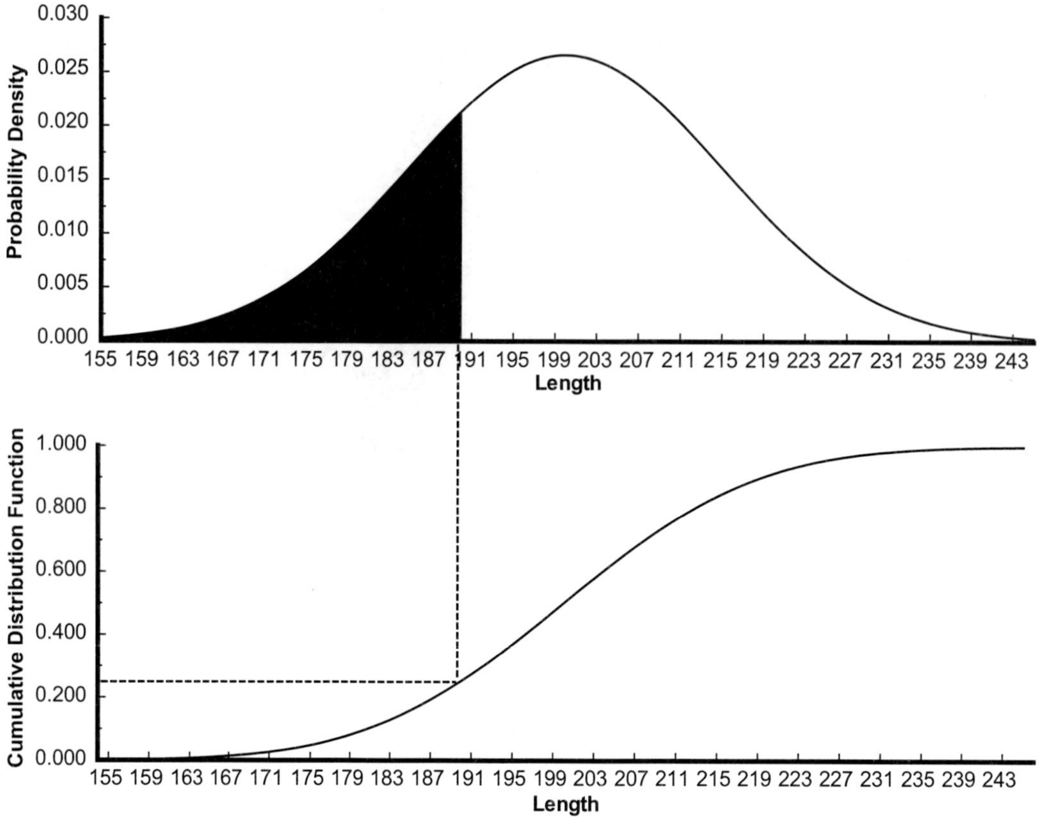

Figure 2.7 *Cumulative distribution function.*

The area of the shaded region of the probability density function in Figure 2.7 is 0.2525. This is the corresponding value of the cumulative distribution function at $x = 190$. Mathematically, the cumulative distribution function is equal to the integral of the probability density function to the left of x.

$$F(x) = \int_{-\infty}^{x} f(\tau)\,d\tau \tag{2.3}$$

Example 2.2: A random variable has the probability density function $f(x) = 0.125x$, where x is valid from 0 to 4. The probability of x being less than or equal to 2 is

$$F(2) = \int_0^2 0.125x \; dx = \left.\frac{0.125x^2}{2}\right|_0^2 = \left.0.0625x^2\right|_0^2 = 0.25$$

Example 2.3: The time to fail for a transistor has the following probability density function. What is the probability of failure before $t = 200$?

$$f(t) = 0.01e^{-0.01t}$$

Solution: The probability of failure before $t = 200$ is

$$P(t < 200) = \int_0^{200} 0.01e^{-0.01t}dt$$

$$= -e^{-0.01t}\Big|_0^{200}$$

$$= -e^{-2} - \left(-e^{0}\right) = 1 - e^{-2} = 0.865$$

Reliability Function

The reliability function is the complement of the cumulative distribution function. If modeling the time to fail, the cumulative distribution function represents the probability of failure, and the reliability function represents the probability of survival. Thus, the cumulative distribution function increases from 0 to 1 as the value of x increases, and the reliability function decreases from 1 to 0 as the value of x increases. This is shown in Figure 2.8.

As seen from Figure 2.8, the probability that the time to fail is greater than 190, the reliability, is 0.7475. The probability that the time to fail is less than 190, the cumulative distribution function, is

$$1 - 0.7475 = 0.2525$$

Mathematically, the reliability function is the integral of the probability density function from x to infinity.

$$R(x) = \int_x^{\infty} f(\tau)d\tau \tag{2.4}$$

Hazard Function

The hazard function is a measure of the tendency to fail; the greater the value of the hazard function, the greater the probability of impending failure. Technically, the hazard function is the probability of failure in the very small time interval, x_0 to $x_{0+\delta}$, given survival until x_0. The hazard function is also known as the instantaneous failure rate. Mathematically, the hazard function is defined as

$$h(x) = \frac{f(x)}{R(x)} \tag{2.5}$$

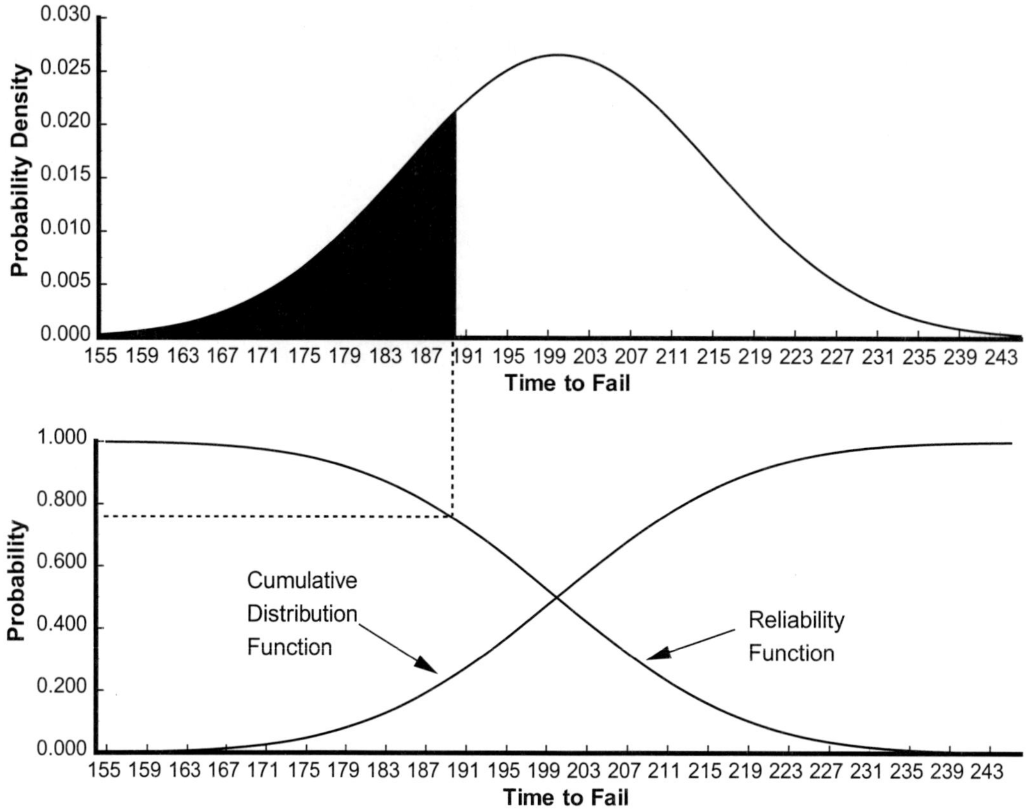

Figure 2.8 *The reliability function.*

Using Eq. 2.5 and Eqs. 2.6 and 2.7, if the hazard function, the reliability function, or the probability density function is known, the remaining two functions can be derived.

$$R(x) = e^{-\int_{-\infty}^{x} h(\tau)d\tau} \tag{2.6}$$

$$f(x) = h(x)e^{-\int_{-\infty}^{x} h(\tau)d\tau} \tag{2.7}$$

Example 2.4: Given the hazard function, $h(x) = 2x$, derive the reliability function and the probability density function.

Solution: The reliability function is

$$R(x) = e^{-\int_{-\infty}^{x} 2x\,dx}$$

$$R(x) = e^{-x^2}$$

The probability density function is

$$f(x) = h(x)R(x) = 2xe^{-x^2}$$

Expectation

Several terms are used to describe distributions. The most common terms are the mean, variance, skewness, and kurtosis. These descriptors are derived from moment-generating functions. Readers with an engineering background may recall that the center of gravity of a shape is

$$cog = \frac{\int_{-\infty}^{\infty} xf(x)\,dx}{\int_{-\infty}^{\infty} f(x)\,dx} \tag{2.8}$$

The mean, or average, of a distribution is its center of gravity. In Eq. 2.8, the denominator is equal to the area below $f(x)$, which is equal to 1 by definition for valid probability distributions. The numerator in Eq. 2.8 is the first moment-generating function about the origin. Thus, the mean of a distribution can be determined from the expression

$$E(x) = \int_{-\infty}^{\infty} xf(x)\,dx \tag{2.9}$$

The second moment-generating function about the origin is

$$E\left(x^2\right) = \int_{-\infty}^{\infty} x^2 f(x)\,dx \tag{2.10}$$

The variance of a distribution is equal to the second moment-generating function about the mean, which is

$$\sigma^2 = \int_{-\infty}^{\infty} x^2 f(x)\,dx - \mu^2 \tag{2.11}$$

The variance is a measure of the dispersion in a distribution. In Figure 2.9, the variance of Distribution A is greater than the variance of Distribution B.

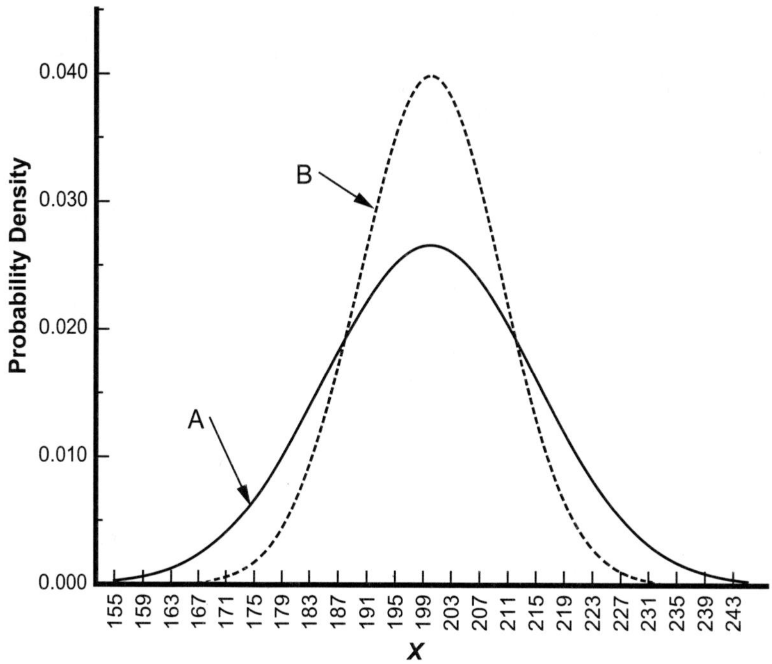

Figure 2.9 *Distribution variance.*

The skewness of a distribution is equal to the third moment-generating function about the mean. If the skewness is positive, the distribution is right skewed. If the skewness is negative, the distribution is left skewed. Right and left skewness are demonstrated in Figure 2.10.

Kurtosis is the fourth moment-generating function about the mean and is a measure of the peakedness of the distribution.

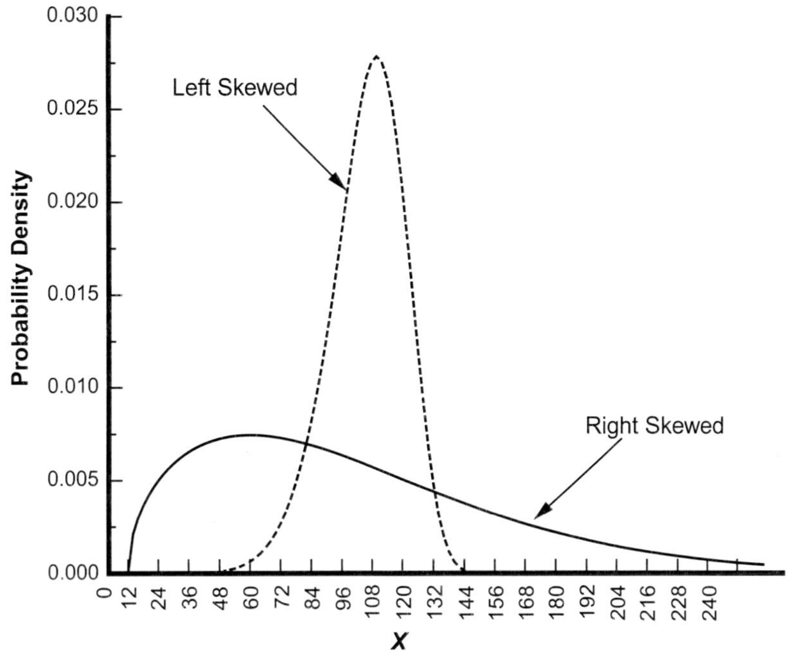

Figure 2.10 *Right and left skewness.*

Summary

A statistical distribution can be described by the following functions:

- Probability density function, $f(x)$
- Cumulative distribution function, $F(x)$
- Reliability function, $R(x)$
- Hazard function, $h(x)$

When one of these functions is defined, the others can be derived. These functions, with characteristics concerning central tendency, spread, and symmetry, provide a description of the population being modeled.

CHAPTER 3

DISTRIBUTIONS

Statistical distributions fall into two categories: (1) modeling distributions, and (2) sampling distributions. Modeling distributions are used to describe data sets and are divided into two classes: (1) continuous distributions, and (2) discrete distributions. Sampling distributions are used to construct confidence intervals and to test hypotheses.

Continuous Modeling Distributions

The four most common reliability modeling distributions are:

1. The Weibull distribution
2. The normal distribution
3. The lognormal distribution
4. The exponential distribution

Each of these distributions is described in detail in the following sections.

Weibull Distribution

The Weibull distribution is one of the most commonly used distributions in reliability to model time to fail, time to repair, and material strength. The Weibull probability density function is

$$f(x) = \frac{\beta}{\theta}\left(\frac{x-\delta}{\theta}\right)^{(\beta-1)} e^{-\left[\left(\frac{x-\delta}{\theta}\right)\right]^{\beta}}, \quad x \geq \delta \tag{3.1}$$

where

β is the shape parameter

θ is the scale parameter

δ is the location parameter

The shape parameter gives the Weibull distribution its flexibility. By changing the value of the shape parameter, the Weibull distribution can model a variety of data. If $\beta = 1$, the Weibull distribution is identical to the exponential distribution. If $\beta = 2$, the Weibull distribution is identical to the Rayleigh distribution. If β is between 3 and 4, the Weibull distribution approximates the normal distribution. The Weibull distribution approximates the lognormal distribution for several values of β. For most populations, more than 50 samples are required to differentiate between the Weibull and lognormal distributions. Figure 3.1 shows a sample of the flexibility of the Weibull distribution.

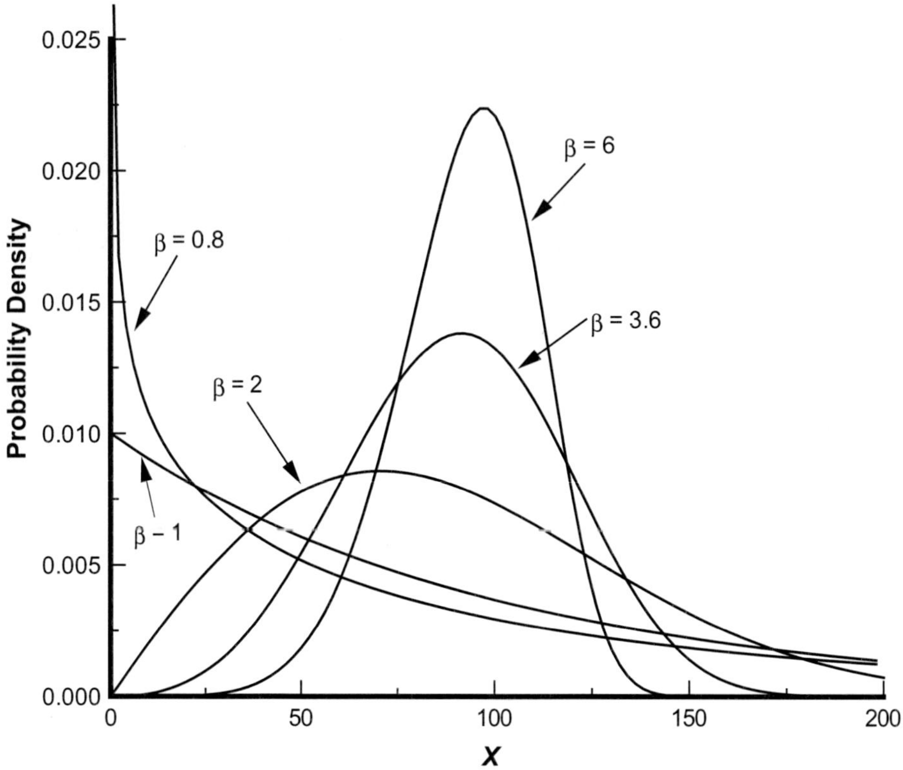

Figure 3.1 *Weibull probability density functions, with $\theta = 100$ and $\delta = 0$.*

The scale parameter determines the range of the distribution. The scale parameter is also known as the characteristic life if the location parameter is equal to 0. If $\delta \neq 0$, the characteristic life is $\theta + \delta$; 63.2% of all values fall below the characteristic life, regardless of the value of the shape parameter. Figure 3.2 shows the effect of the scale parameter of the probability density function.

The location parameter is used to define a failure-free zone. The probability of failure when $x < \delta$ is 0. When $\delta > 0$, there is a period when no failures can occur. When $\delta < 0$, failures have occurred before time equals 0. At first, this seems ridiculous; however, a negative location parameter is caused by shipping failed units, failures during transportation, and shelf-life failures. Generally, the location parameter is assumed to be zero. Figure 3.3 shows the effect of the location parameter.

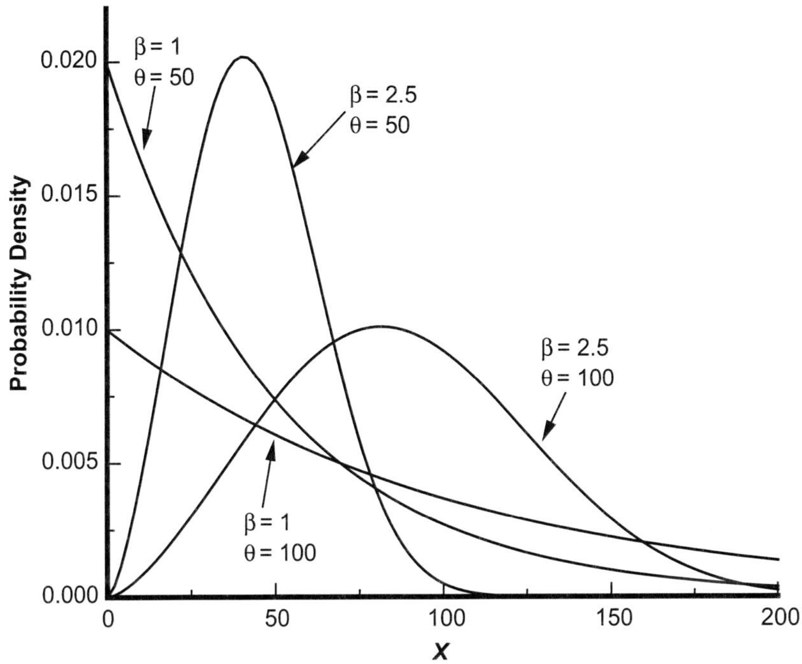

Figure 3.2 *Effect of the Weibull scale parameter.*

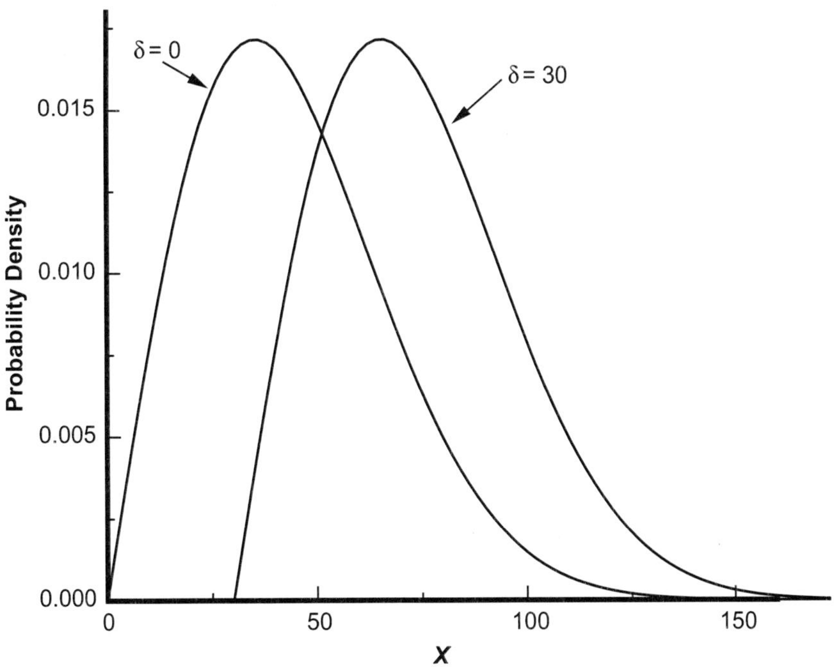

Figure 3.3 *Effect of the Weibull location parameter.*

The Weibull hazard function is determined by the value of the shape parameter,

$$h(x) = \frac{\beta}{\theta}\left(\frac{x-\delta}{\theta}\right)^{(\beta-1)} \tag{3.2}$$

When $\beta < 1$, the hazard function is decreasing; this is known as the infant mortality period. When $\beta = 1$, the failure rate is constant. When $\beta > 1$, the failure rate is increasing; this is known as the wearout period. Figure 3.4 shows the Weibull hazard function.

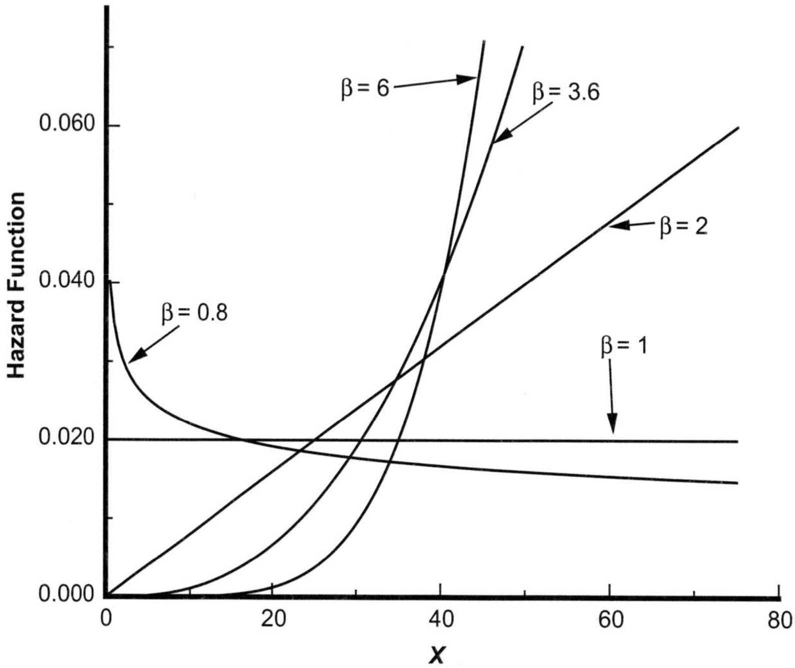

Figure 3.4 *The Weibull hazard function.*

When $\beta < 1$, the hazard function decreases more steeply as β decreases. When $\beta > 1$, the rate of increase for the hazard function increases as β increases. As seen from Figure 3.4, when $\beta = 2$, the hazard function increases linearly.

The Weibull reliability and cumulative distribution functions are, respectively,

$$R(x) = e^{-\left(\frac{x-\delta}{\theta}\right)^{\beta}} \tag{3.3}$$

and

$$F(x) = 1 - e^{-\left(\frac{x-\delta}{\theta}\right)^{\beta}} \tag{3.4}$$

Example 3.1: The time to fail for a flexible membrane follows the Weibull distribution with $\beta = 2$ and $\theta = 300$ months. What is the reliability at 200 months? After how many months is 90% reliability achieved?

Solution: After 200 months, the reliability of the flexible membrane is

$$R(200) = e^{\left[-\left(\frac{200}{300}\right)^{2}\right]} = 0.6412$$

By manipulating the expression for reliability, 90% reliability is achieved after

$$t = \theta(-\ln R)^{\left(\frac{1}{\beta}\right)}$$

$$t = 300\sqrt{-\ln(0.9)} = 97.38 \text{ months}$$

The mean and variance of the Weibull distribution are computed using the gamma distribution, which is given in Table A.1 of Appendix A.* The mean of the Weibull distribution is

$$\mu = \theta\Gamma\left(1 + \frac{1}{\beta}\right) \tag{3.5}$$

The mean of the Weibull distribution is equal to the characteristic life if the shape parameter is equal to 1. Figure 3.5 shows the mean as a function of the shape parameter.

The variance of the Weibull distribution is

$$\sigma^2 = \theta^2\left[\Gamma\left(1 + \frac{2}{\beta}\right) - \Gamma^2\left(1 + \frac{1}{\beta}\right)\right] \tag{3.6}$$

The variance of the Weibull distribution decreases as the value of the shape parameter increases, as shown in Figure 3.6.

* The gamma function is available in Microsoft® Excel. The function is =EXP(GAMMALN(x)). The function GAMMALN(x) returns the natural logarithm of the gamma function; the function EXP(x) transforms the expression into the gamma function.

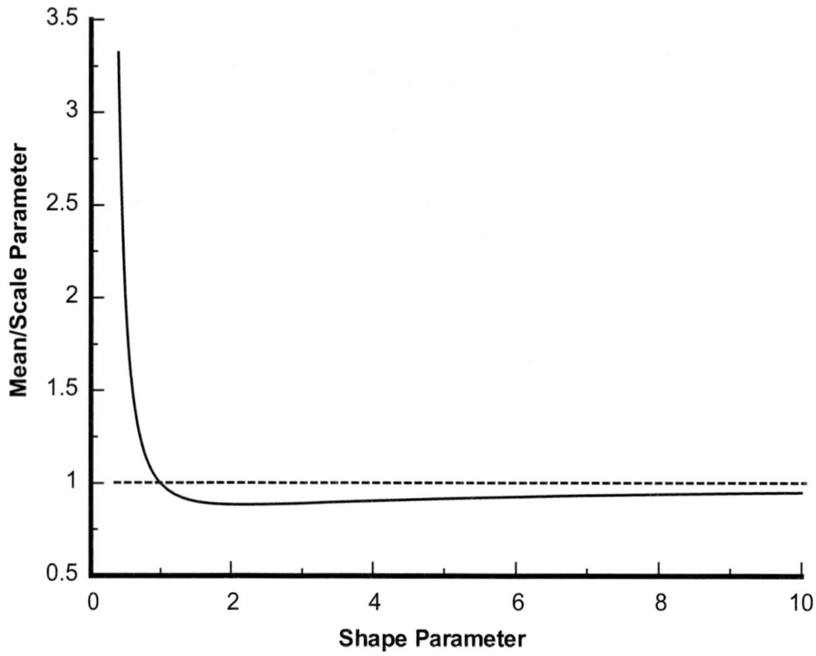

Figure 3.5 *The Weibull mean.*

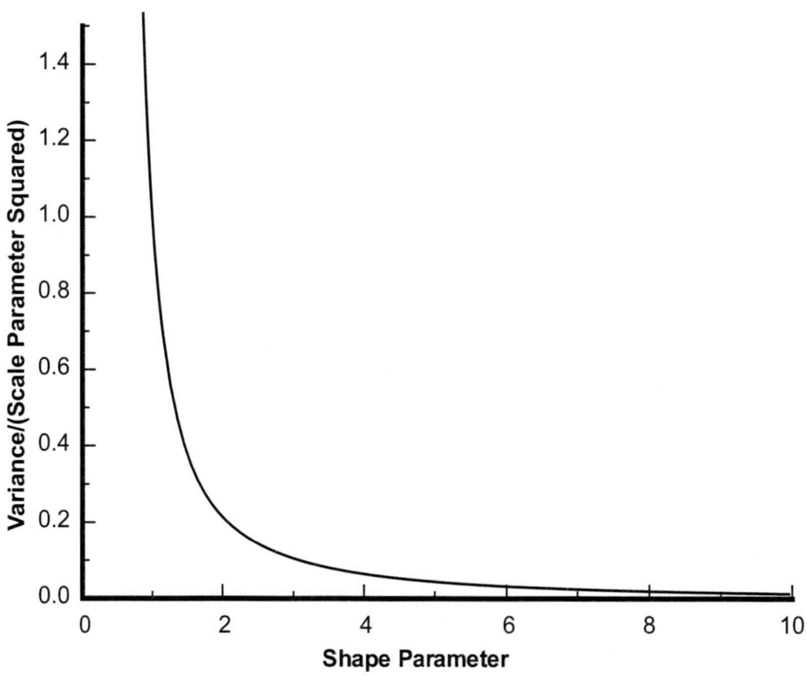

Figure 3.6 *The Weibull variance.*

The reliability measure of many components is the mean time to fail. Consumer electronics companies often advertise the mean time to fail of the products on television. The mean time to fail is a deceptive measure because the variance of the time-to-fail distribution is not considered. To achieve the same reliability with a larger variance requires a larger mean.

Consider two components, A and B. Component A has a mean time to fail of 4,645 hours, and Component B has a mean time to fail of 300 hours. If both components sell for the same price, which component should be used to maximize reliability at 100 hours?

This question cannot be answered without knowing more information about the distribution of the time to fail. Component A has a mean of 4,645 hours and a Weibull time-to-fail distribution with a shape parameter of 0.8. Using the mean and the shape parameter, the scale parameter of Component A can be computed to be 4,100 hours. The reliability at 100 hours is

$$R(100) = e^{-\left(\frac{100}{4100}\right)^{0.8}} = 0.95$$

Component B has a mean of 300 hours and a Weibull time-to-fail distribution with a shape parameter of 3. Using the mean and the shape parameter, the scale parameter of Component B can be computed to be 336 hours. The reliability at 100 hours is

$$R(100) = e^{-\left(\frac{100}{336}\right)^{3}} = 0.974$$

Although the mean of Component A is more than 10 times as large as the mean of Component B, the reliability of Component B is greater than the reliability of Component A at 100 hours. Continuing with this example, if the reliability at 1,000 hours is to be maximized, Component A has a reliability of 0.723, and Component B has a reliability of approximately 0.

Normal Distribution

Whenever several random variables are added together, the resulting sum tends to normal, regardless of the distribution of the variables being added. Mathematically, if

$$y = x_1 + x_2 + x_3 + \ldots + x_n \tag{3.7}$$

then the distribution of y becomes normal as n increases. If the random variables being summed are independent, the mean and variance of y are, respectively,

$$\mu_y = \mu_{x1} + \mu_{x2} + \mu_{x3} + \ldots + \mu_{xn} \tag{3.8}$$

and

$$\sigma_y^2 = \sigma_{x1}^2 + \sigma_{x2}^2 + \sigma_{x3}^2 + \ldots + \sigma_{xn}^2 \tag{3.9}$$

When several random variables are averaged, the resulting average tends to normal, regardless of the distribution of the variables being averaged. Mathematically, if

$$y = \frac{x_1 + x_2 + x_3 + \ldots + x_n}{n} \tag{3.10}$$

then the distribution of y becomes normal as n increases. If the random variables being averaged have the same mean and variance, then the mean of y is equal to the mean of the individual variables being averaged, and the variance of y is

$$\sigma_y^2 = \frac{\sigma^2}{n} \tag{3.11}$$

where σ^2 is the variance of the individual variables being averaged.

The tendency of sums and averages to become normally distributed as the number of variables being summed or averaged becomes large is known as the central limit theorem or the theory of large numbers. For distributions with little skewness, summing or averaging as few as three or four variables will result in a normal distribution. For highly skewed distributions, more than thirty variables may have to be summed or averaged to obtain a normal distribution.

The normal probability density function is

$$f(x) = \frac{1}{\sigma\sqrt{2\Pi}} e^{-\frac{1}{2}\left(\frac{x-\mu}{\sigma}\right)^2}, \quad -\infty < x < \infty \tag{3.12}$$

where μ is the mean and σ is the standard deviation.

The normal probability density function is not skewed and is shown in Figure 3.7.

The density function shown in Figure 3.7 is the standard normal probability density function. The standard normal probability density function has a mean of 0 and a standard deviation of 1. The normal probability density function cannot be integrated implicitly. Because of this, historically a transformation to the standard normal distribution is made, and the normal cumulative distribution function or reliability function is read from a table. Table A.2 in Appendix A gives the standard normal reliability function. If x is a normal random variable, it can be transformed to standard normal using the expression

$$z = \frac{x - \mu}{\sigma} \tag{3.13}$$

Example 3.2: The tensile strength of a metal extrusion is normally distributed with a mean of 300 and a standard deviation of 5. What percentage of extrusions has a strength greater than 310? What percentage of extrusions has a strength less than 295? What percentage of extrusions has a strength between 295 and 310?

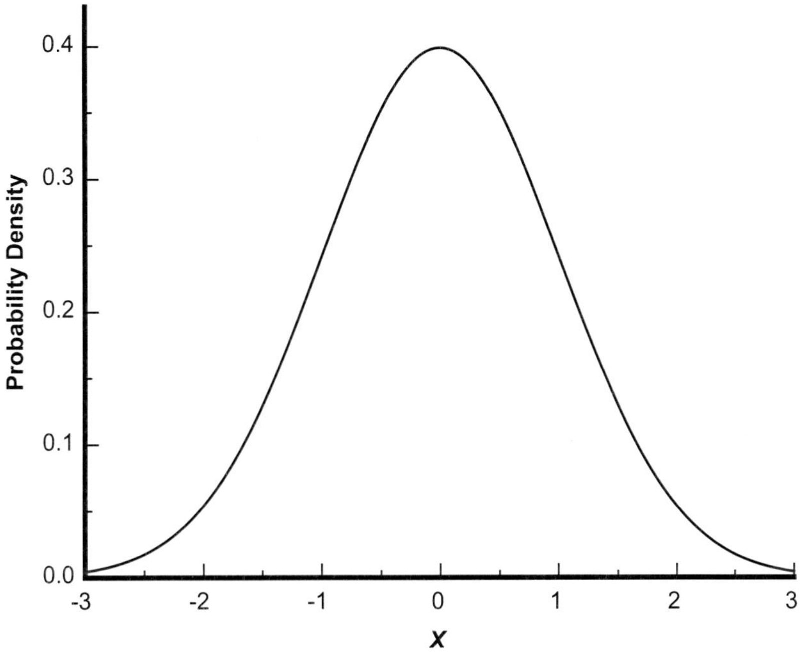

Figure 3.7 *The standard normal probability density function.*

Solution: The shaded area in the top graph in Figure 3.8 represents the probability of an extrusion being greater than 310. The shaded area in the bottom graph in Figure 3.8 represents the area under the standard normal distribution to the right of $z = 2$, which is the same as the probability of an extrusion being greater than 310. Transforming to standard normal,

$$z = \frac{310 - 300}{5} = 2$$

To determine the area under the standard normal probability density function to the right of $z = 2$, look up $z = 2$ in Table A.1 of Appendix A, which is 0.0228. The percentage of extrusions with strength greater than 310 is 0.0228.

The shaded area in the top graph in Figure 3.9 represents the probability of an extrusion being less than 295. The shaded area in the bottom graph in Figure 3.9 represents the area under the standard normal distribution to the left of $z = -1$, which is the same as the probability of an extrusion being less than 295. Transforming to standard normal,

$$z = \frac{295 - 300}{5} = -1$$

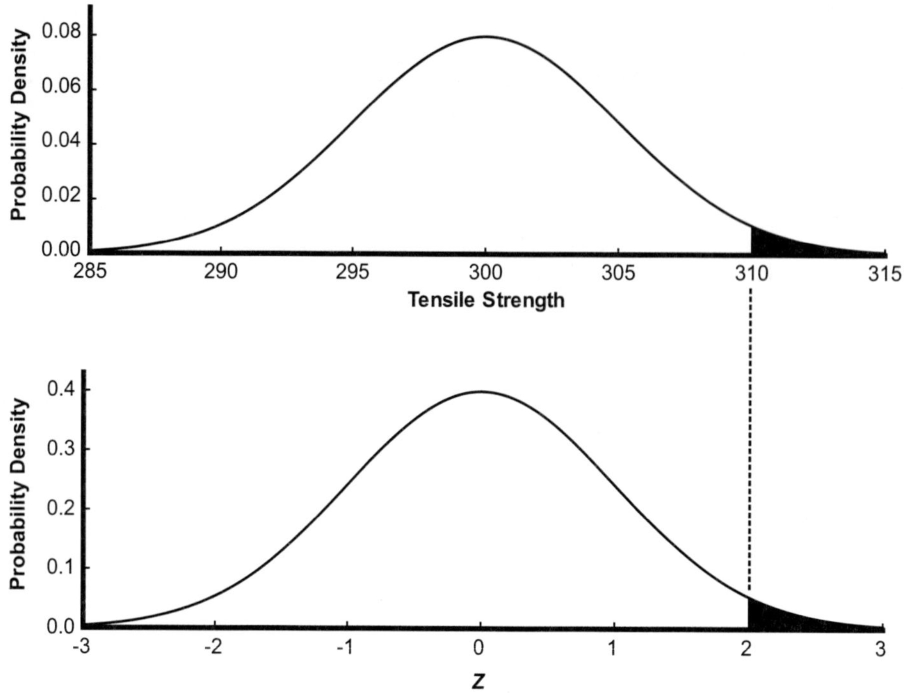

Figure 3.8 *Probability of strength being greater than 310.*

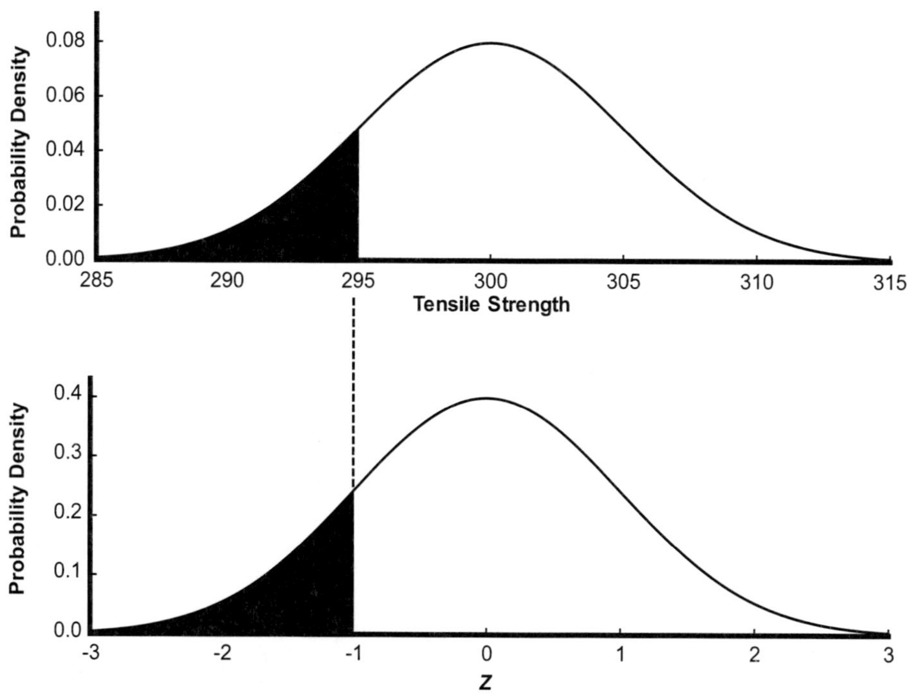

Figure 3.9 *Probability of strength being less than 295.*

From Table A.1 of Appendix A, the area to the left of $z = -1$ is equal to 0.1587. The probability of the strength being between 295 and 310 is

$$1 - 0.0228 - 0.1587 = 0.8185$$

The transformation to standard normal is no longer required. Electronic spreadsheets have functions to evaluate the normal probability density function. The area to the left of x under the normal probability density function with a mean of μ and a standard deviation of σ is found using the Microsoft Excel expression

$$\text{=NORMDIST}\left(x, \mu, \sigma, 1\right)$$

The inverse of this function is also available in Microsoft Excel. Given a normal distribution with a mean of μ and a standard deviation of σ, the value of y having an area of x under the probability density function to the left of y is

$$\text{=NORMINV}\left(x, 0, 1\right)$$

Example 3.3: Repeat Example 3.2 using Microsoft Excel.

Solution: The percentage of extrusions having a strength less than 310 is found using the expression

$$\text{=NORMDIST}\left(310, 300, 5, 1\right)$$

The percentage of extrusions having a strength greater than 310 is found using the expression

$$\text{=1-NORMDIST}\left(310, 300, 5, 1\right)$$

The percentage of extrusions having a strength less than 295 is found using the expression

$$\text{=NORMDIST}\left(295, 300, 5, 1\right)$$

The percentage of extrusions having a strength between 295 and 310 is found using the expression

$$\text{=NORMDIST}\left(310, 300, 5, 1\right) \text{- NORMDIST}\left(295, 300, 5, 1\right)$$

Example 3.4: A type of battery is produced with an average voltage of 60 with a standard deviation of 4 volts. If nine batteries are selected at random, what is the probability that the total voltage of the nine batteries is greater than 530? What is the probability that the average voltage of the nine batteries is less than 62?

Solution: The expected total of the voltage of nine batteries is 540. The expected standard deviation of the voltage of the total of nine batteries is

$$\sigma = \frac{(9)(4)}{\sqrt{9}} = 12$$

Transforming to standard normal,

$$z = \frac{530 - 540}{12} = -0.833$$

From Table A.1 of Appendix A, the area to the left of $z = -0.833$ is 0.2024. The area to the right of $z = -0.833$ is $1 - 0.2024 = 0.7976$. This value can also be computed using the Microsoft Excel function

=1-NORMDIST(530,540,12,1)

The probability density function of the voltage of the individual batteries and of the average of nine batteries is shown in Figure 3.10. The distribution of the averages has less variance because the standard deviation of the averages is equal to the standard deviation of the individuals divided by the square root of the sample size.

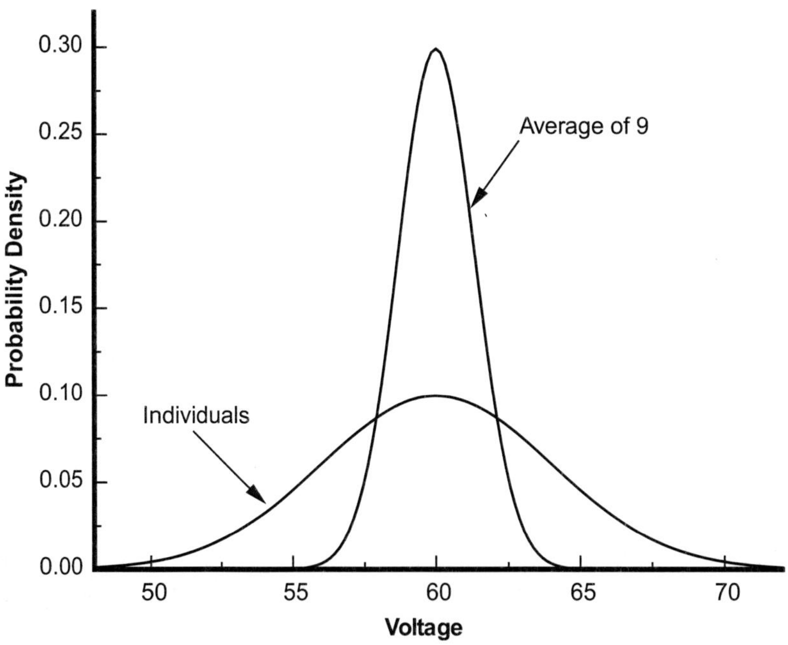

Figure 3.10 *Distribution of average voltage.*

The probability of the average voltage of nine batteries being less than 62 is equal to the probability of a standard normal variable being less than $z = 1.5$. This is shown in the expression

$$z = \frac{62 - 60}{4/\sqrt{9}} = 1.5$$

From Table A.1 of Appendix A, the area under the standard normal curve to the left of $z = 1.5$ is 0.9332, which is the probability of the average voltage of nine batteries being less than 62. This value can also be computed using the Microsoft Excel function

=NORMDIST(62,60,1.333,1)

Example 3.5: A type of battery is produced with an average voltage of 12 with a standard deviation of 1.4 volts. Ninety percent of the batteries have a voltage greater than what value?

Solution: From Table A.1 in Appendix A, 90% of the area under the standard normal probability density function is to the right of $z = 1.282$. Transforming from standard normal to a normal distribution with a mean of 12 and a standard deviation of 1.4 gives

$$1.282 = \frac{x - 12}{1.4}$$

$$x = 10.21$$

The voltage of 90% of the batteries is greater than 10.21. The value can be found using the Microsoft Excel function

=NORMINV(0.1,12,1.4)

The normal hazard function is

$$h(x) = \frac{\phi\left(\frac{x-\mu}{\sigma}\right)}{\sigma\left[1 - \Phi\left(\frac{x-\mu}{\sigma}\right)\right]} \tag{3.14}$$

where $\varphi(x)$ is the standard normal probability density function and $\Phi(x)$ is the standard normal cumulative distribution function.

The hazard function for the normal distribution is monotonically increasing, as shown in Figure 3.11.

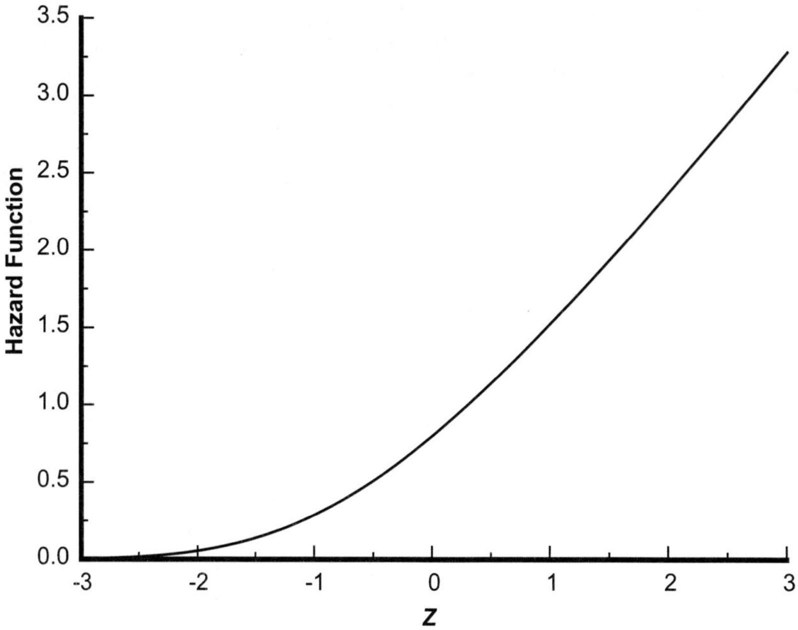

Figure 3.11 *Normal hazard function.*

Lognormal Distribution

If a data set is known to follow a lognormal distribution, transforming the data by taking a logarithm yields a data set that is normally distributed, as shown in Table 3.1.

TABLE 3.1
TRANSFORMATION OF LOGNORMAL DATA

Lognormal Data	Normal Data
12	ln(12)
16	ln(16)
28	ln(28)
48	ln(48)
87	ln(87)
143	ln(143)

The most common transformation is made by taking the natural logarithm, but any base logarithm, such as base 10 or base 2, also yields a normal distribution. The remaining discussion will use the natural logarithm denoted as "ln."

When random variables are summed, as the sample size increases, the distribution of the sum becomes a normal distribution, regardless of the distribution of the individuals. Because lognormal random variables are transformed to normal random variables by taking the logarithm, when random variables are multiplied, as the sample size increases, the distribution of the product becomes a lognormal distribution, regardless of the distribution of the individuals. This is because the logarithm of the product of several variables is equal to the sum of the logarithms of the individuals. This is shown as

$$y = x_1 x_2 x_3$$

$$\ln y = \ln x_1 + \ln x_2 + \ln x_3$$

The lognormal probability density function is

$$f(x) = \frac{1}{x\sigma\sqrt{2\Pi}} e^{-\frac{1}{2}\left(\frac{\ln x - \mu}{\sigma}\right)^2}, \quad x > 0 \tag{3.15}$$

where μ is the location parameter or log mean, and σ is the scale parameter or log standard deviation. The location parameter is the mean of the data set after transformation by taking the logarithm, and the scale parameter is the standard deviation of the data set after transformation.

The lognormal distribution takes on several shapes, depending on the value of the shape parameter. The lognormal distribution is skewed right, and the skewness increases as the value of σ increases, as shown in Figure 3.12.

The lognormal cumulative distribution and reliability functions are, respectively,

$$F(x) = \Phi\left(\frac{\ln x - \mu}{\sigma}\right)$$

$$R(x) = 1 - \Phi\left(\frac{\ln x - \mu}{\sigma}\right)$$

where $\Phi(x)$ is the standard normal cumulative distribution function.

Example 3.6: The following data are the time to fail for four light bulbs and are known to have a lognormal distribution. What is the reliability at 100 hours?

115 hours	155 hours
183 hours	217 hours

Solution: Table 3.2 shows this data and the transformation to normal.

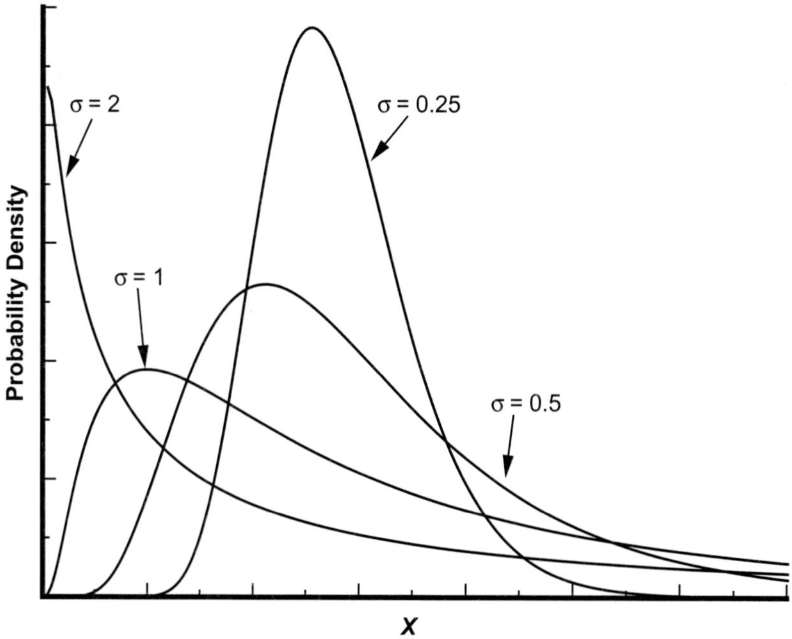

Figure 3.12 Lognormal probability density function.

TABLE 3.2
TRANSFORMATION TO NORMAL

Time to Fail	ln(Time to Fail)
115	4.7449
155	5.0434
183	5.2095
217	5.3799

The parameters of the lognormal distribution are found by computing the mean and the standard deviation of the transformed data in the second column of Table 3.2. The mean of the transformed data is

$$\mu = \frac{4.7449 + 5.0434 + 5.2095 + 5.3799}{4} = 5.0944$$

The sum of the second column of Table 3.2 is 20.3777. If each value in the second column of Table 3.2 is squared, the sum of the squared values is 104.0326. Using these values, the sample standard deviation of the values in the second column of Table 3.2 is

$$s = \sqrt{\frac{n\sum_{i=1}^{n} x_i^2 - \left(\sum_{i=1}^{n} x\right)^2}{n(n-1)}} = \sqrt{\frac{4(104.0326) - 20.3777^2}{4(4-1)}} = 0.2708$$

An easier method of computing the standard deviation is to use the Microsoft Excel function

=stdev()

The reliability at 100 hours is

$$R(100) = 1 - \Phi\left(\frac{\ln(100) - 5.0944}{0.2708}\right) = 1 - \Phi(-1.807)$$

From Table A.1 of Appendix A, the standard normal cumulative distribution function at $z = -1.807$ is 0.0354. Thus, the reliability at 100 hours is

$$R(100) = 1 - 0.0354 = 0.9646$$

The reliability at 100 hours can also be computed using the Microsoft Excel function

=1-NORMDIST(LN(100),5.0944,0.2708,1)

The location parameter, or log mean, is often mistaken for the mean of the lognormal distribution. The mean of the lognormal distribution can be computed from its parameters

$$\text{mean} = e^{\left(\mu + \frac{\sigma^2}{2}\right)}$$

The variance of the lognormal distribution is

$$\text{variance} = \left(e^{\left(2\mu + \sigma^2\right)}\right)\left(e^{\sigma^2} - 1\right)$$

The lognormal hazard function has a unique behavior; it increases initially, then decreases, and eventually approaches zero. This means that items with a lognormal distribution have a higher chance of failing as they age for some period of time, but after survival to a specific age, the probability of failure decreases as time increases. Figure 3.13 shows the lognormal hazard function.

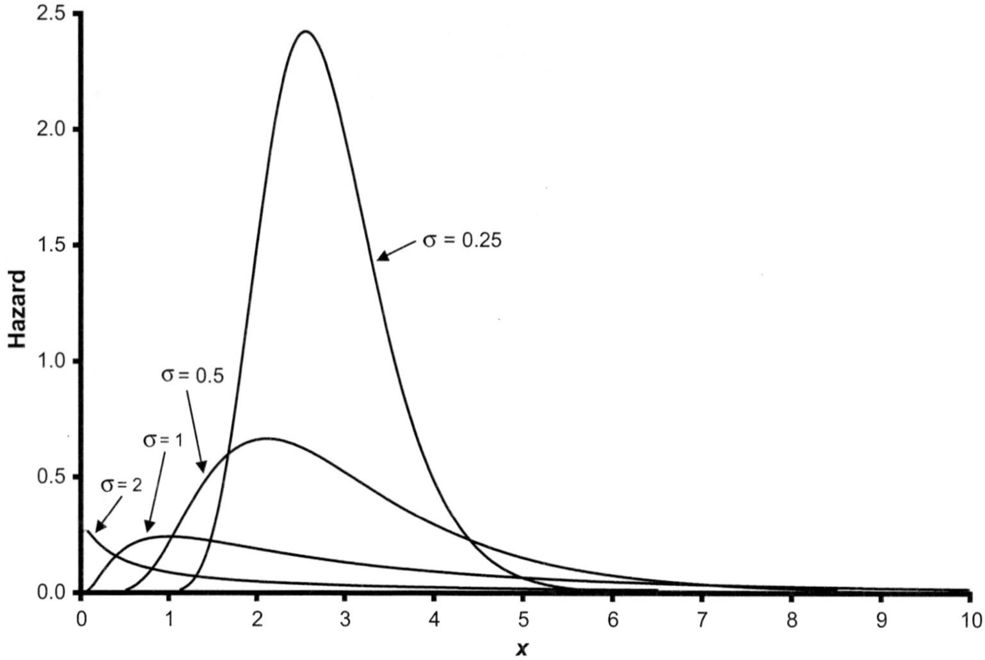

Figure 3.13 *The lognormal hazard function.*

In Figure 3.13, when $\sigma = 2$, the hazard function increases so quickly that it cannot be seen on the graph.

Exponential Distribution

The exponential distribution is used to model items with a constant failure rate, usually electronics. The exponential distribution is closely related to the Poisson distribution. If a random variable x is exponentially distributed, then the reciprocal of x, $y = \frac{1}{x}$, follows a Poisson distribution. Likewise, if x is Poisson distributed, then $y = \frac{1}{x}$ is exponentially distributed. Because of this behavior, the exponential distribution is usually used to model the mean time between occurrences, such as arrivals or failures, and the Poisson distribution is used to model occurrences per interval, such as arrivals, failures, or defects.

The exponential probability density function is

$$f(x) = \lambda e^{-\lambda x}, \quad x > 0 \tag{3.16}$$

where λ is the failure rate.

The exponential probability density function is also written as

$$f(x) = \frac{1}{\theta} e^{-\frac{x}{\theta}}, \quad x > 0 \tag{3.17}$$

where θ is the mean.

From these equations, it can be seen that $\lambda = \frac{1}{\theta}$. The variance of the exponential distribution is equal to the mean squared,

$$\sigma^2 = \theta^2 = \frac{1}{\lambda^2} \tag{3.18}$$

Figure 3.14 shows the exponential probability density function.

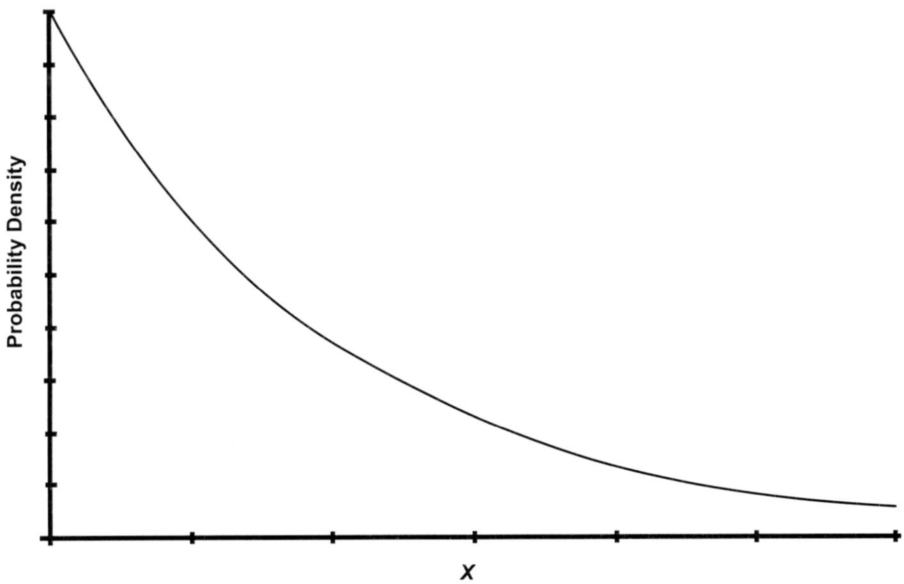

Figure 3.14 *Exponential probability density function.*

The exponential reliability function is

$$R(x) = e^{-\frac{x}{\theta}} = e^{-\lambda x}, \quad x > 0 \tag{3.19}$$

The exponential hazard function is

$$h(x) = \frac{1}{\theta} = \lambda \tag{3.20}$$

Figure 3.15 shows the exponential hazard function.

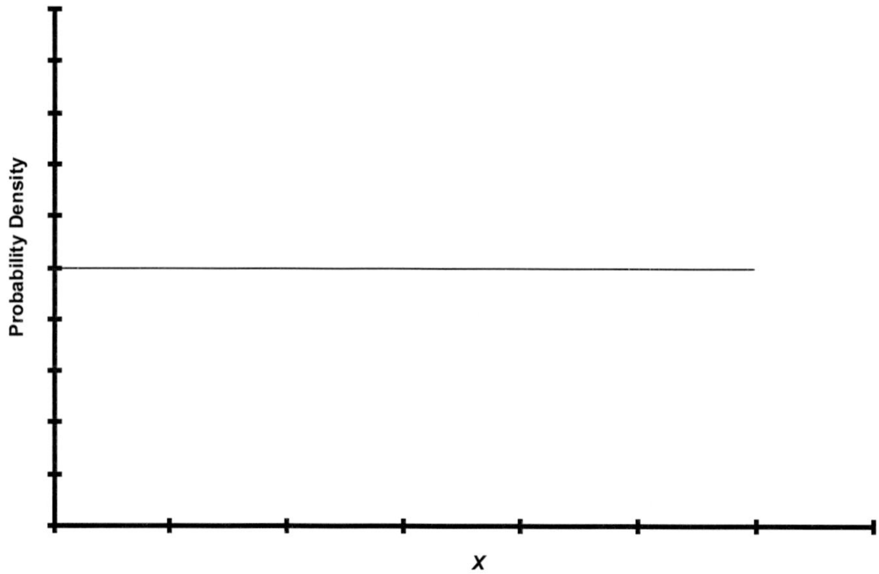

Figure 3.15 *Exponential hazard function.*

Example 3.7: A resistor has a constant failure rate of 0.04 per hour. What is the reliability of the resistor at 100 hours? If 100 resistors are tested, how many would be expected to be in a failed state after 25 hours?

Solution: The reliability at 100 hours is

$$R(100) = e^{-0.04(100)} = 0.0183$$

The probability of failing before 25 hours is given by the cumulative distribution function, which is equal to 1 minus the reliability function,

$$F(25) = 1 - R(25) = 1 - e^{0.04(25)} = 1 - 0.368 = 0.632$$

The expected number of resistors in a failed state is $100(0.632) = 63.2$.

The exponential distribution is characterized by its hazard function, which is constant. Because of this, the exponential distribution exhibits a lack of memory. That is, the probability of survival for a time interval, given survival to the beginning of the interval, is dependent only on the length of the interval and not on the time of the start of the interval. For example, consider an item that has a mean time to fail of 150 hours that is exponentially distributed. The probability of surviving through the interval 0 to 20 hours is

$$R(20) = e^{-\frac{20}{150}} = 0.8751$$

The probability of surviving the interval 100 to 120 is equal to

$$R(120, \text{ given survival to } t = 100) = \frac{R(120)}{R(100)} = \frac{e^{-\frac{120}{150}}}{e^{-\frac{100}{150}}} = \frac{0.4493}{0.5134} = 0.8751$$

The mean of the exponential distribution is estimated using the expression

$$\theta = \frac{T}{r} \tag{3.21}$$

where T represents the total test time for all items, both failed and unfailed.

A $(1-\alpha)\%$ confidence limit for θ is

$$\frac{2T}{\chi^2_{\alpha,\, 2r+2}} \leq \theta \leq \frac{2T}{\chi^2_{1-\alpha,\, 2r}} \tag{3.22}$$

Equation 3.22 assumes time censoring. If the data are failure censored, the critical value of the chi-square statistic has $2r$ degrees of freedom instead of $2r + 2$,

$$\frac{2T}{\chi^2_{\alpha,\, 2r}} \leq \theta \leq \frac{2T}{\chi^2_{1-\alpha,\, 2r}} \tag{3.23}$$

Confidence limits for reliability can be found by substituting the lower and upper limits for θ into the exponential reliability function.

Example 3.8: Five items were tested with failures after 43 hours, 57 hours, and 80 hours. The remaining two items were removed from testing after 80 hours without failing. Find the 90% confidence interval for reliability at 130 hours.

Solution: Because the items that did not fail were removed from testing at the same time as the last failure, this is failure censoring. To determine the confidence limits for reliability, the confidence limits for the mean must be determined first. The critical value of the chi-square distribution with $\alpha = 0.05$ and 6 degrees of freedom is 2.592. This is found from Appendix A or from the Microsoft Excel expression

$$= \text{CHIINV}(0.05,6)$$

The critical value of the chi-square distribution with $\alpha = 0.95$ and 6 degrees of freedom is 1.635. This is found from Appendix A or from the Microsoft Excel expression

$$=\text{CHIINV}(0.95,6)$$

The 90% confidence interval for the mean is

$$\frac{2(43+57+80+80+80)}{12.592} \le \theta \le \frac{2(43+57+80+80+80)}{1.635}$$

$$54.0 \le \theta \le 415.9$$

The 90% confidence interval for reliability is

$$e^{-\frac{130}{54.0}} \le e^{-\frac{130}{415.9}}$$

$$0.090 \le 0.732$$

Example 3.9: Five items were tested with failures after 43 hours, 57 hours, and 80 hours. The remaining two items survived for 100 hours without failing. Find the 90% confidence interval for reliability at 130 hours.

Solution: Because the items that did not fail were not removed from testing at the same time as the last failure, this is time censoring. The critical value of the chi-square distribution for the lower limit has $2r + 2$ degrees of freedom, and the critical value for the upper limit has $2r$ degrees of freedom. The critical value of the chi-square distribution with $\alpha = 0.05$ and 8 degrees of freedom is 15.507. The critical value of the chi-square distribution with $\alpha = 0.95$ and 6 degrees of freedom is 1.635. The 90% confidence interval for the mean is

$$\frac{2(43+57+80+100+100)}{15.507} \le \theta \le \frac{2(43+57+80+100+100)}{1.635}$$

$$49.01 \le \theta \le 464.8$$

The 90% confidence interval for reliability is

$$e^{-\frac{130}{49.01}} \le e^{-\frac{130}{464.8}}$$

$$0.0705 \le 0.756$$

Discrete Modeling Distributions

The Poisson, binomial, hypergeometric, and geometric distributions are used to model discrete data. Time to fail is continuous data, but some situations call for discrete data, such as tests that are either pass or fail.

Poisson Distribution

The Poisson distribution is used to model rates, such as rabbits per acre, defects per unit, or arrivals per hour. The Poisson distribution is closely related to the exponential distribution. If x is a Poisson-distributed random variable, then $\frac{1}{x}$ is an exponential random variable. If x is an exponential random variable, then $\frac{1}{x}$ is a Poisson random variable. For a random variable to be Poisson distributed, the probability of an occurrence in an interval must be proportional to the length of the interval, and the number of occurrences per interval must be independent.

The Poisson probability density function is

$$p(x,\mu) = \frac{e^{-\mu}\mu^{x}}{x!} \tag{3.24}$$

The term $p(x,\mu)$ represents the probability of exactly x occurrences in an interval having an average of μ occurrences. The mean and variance of the Poisson distribution are both equal to μ. The Poisson cumulative distribution function is simply the sum of the Poisson probability density function from 0 to x,

$$P(x,\mu) = \sum_{i=0}^{x} \frac{e^{-\mu}\mu^{i}}{i!} \tag{3.25}$$

The cumulative Poisson distribution has been computed in Appendix A to eliminate the need for tedious calculations, but most people prefer to use the Microsoft Excel formula

=POISSON(x,μ,1)

Example 3.10: A complex software system averages 7 errors per 5,000 lines of code. What is the probability of exactly 2 errors in 5,000 lines of randomly selected lines of code?

Solution: The probability of exactly 2 errors in 5,000 lines of randomly selected lines of code is

$$p(2,7) = \frac{e^{-7}7^{2}}{2!} = 0.022$$

Example 3.11: A complex software system averages 7 errors per 5,000 lines of code. What is the probability of exactly 3 errors in 15,000 lines of randomly selected lines of code?

The average number of errors in 15,000 lines of code is

$$\mu = \left(\frac{7}{5,000}\right)(15,000) = 21$$

Solution: The probability of exactly 3 errors in 15,000 lines of randomly selected lines of code is

$$p(3,21) = \frac{e^{-21}21^3}{3!} = 0.00000117$$

Example 3.12: A complex software system averages 6 errors per 5,000 lines of code. What is the probability of fewer than 3 errors in 2,500 lines of randomly selected lines of code? What is the probability of more than 2 errors in 2,500 lines of randomly selected lines of code?

Solution: The average number of errors in 2,500 lines of code is $\mu = 3$. The probability of fewer than 3 defects is equal to the probability of exactly 0 defects plus the probability of exactly 1 defect plus the probability of exactly 2 defects. Entering the cumulative Poisson table in Appendix A with $r = 2$ and $\mu = 3$ gives the probability of 2 or fewer defects, which is 0.4232. This value can also be computed manually. The same solution is found using the Microsoft Excel formula

$$\text{=POISSON}(2,3,1)$$

The “1” at the end of this formula gives the cumulative Poisson.

The probability of more than 2 errors is equal to the probability of exactly 3 plus the probability of exactly 4 plus the probability of exactly 5, and so forth. A simpler approach is to consider that the probability of more than 2 errors is equal to 1 minus the probability of 2 or fewer errors. Thus, the probability of more than 2 errors is

$$1 - 0.4232 = 0.5768$$

Binomial Distribution

The binomial distribution is used to model situations having only two possible outcomes, usually labeled as success or failure. For a random variable to follow a binomial distribution, the number of trials must be fixed, and the probability of success must be equal for all trials. The binomial probability density function is

$$p(x,n,p) = \binom{n}{x} p^x (1-p)^{n-x} \tag{3.26}$$

where $p(x,n,p)$ is the probability of exactly x successes in n trials with a probability of success equal to p on each trial. Note that

$$\binom{n}{x} = \frac{n!}{x!(n-x)!} \tag{3.27}$$

This notation is referred to as "n choose x" and is equal to the number of combinations of size x made from n possibilities. This function is found on most calculators.

The binomial cumulative distribution function is

$$P(x,n,p) = \sum_{i=0}^{x} \binom{n}{i} p^i (1-p)^{n-i} \tag{3.28}$$

where $P(x,n,p)$ is the probability of exactly x or fewer successes in n trials with a probability of success equal to p on each trial.

The mean and variance of the binomial distribution are, respectively,

$$\mu = np \tag{3.29}$$

$$\sigma^2 = np(1-p) \tag{3.30}$$

Example 3.13: The probability of a salesman making a successful sales call is 0.2, and 8 sales calls are made in a day. What is the probability of making exactly 2 successful sales calls in a day? What is the probability of making more than 2 successful sales calls in a day?

Solution: The probability of exactly 2 successes in 8 trials is

$$p(2,8,0.2) = \binom{8}{2} 0.2^2 (1-0.2)^{8-2} = 0.2936$$

This solution is found using the Microsoft Excel function

=BINOMDIST(2,8,0.2,0)

The probability of more than 2 successes is equal to 1 minus the probability of 2 or fewer successes. The binomial probability density function can be used to compute the probability of exactly 0 successes and the probability of exactly 1 success, but it is easier to use the Microsoft Excel function

=BINOMDIST(2,8,0.2,1)

The "1" at the end of this function returns the cumulative binomial distribution. The probability of 2 or fewer successes is 0.7969. The probability of more than 2 successes is

$$1 - 0.7969 = 0.2031$$

Before electronic spreadsheets were common, the Poisson distribution was used to approximate the binomial distribution because Poisson tables were more accommodating than binomial tables. This approximation is now useless; why approximate a value when you can get an exact answer? The requirement for a valid approximation is that p must be small and n must be large. The approximation is done by using np as the mean of the Poisson distribution.

Hypergeometric Distribution

The hypergeometric distribution is similar to the binomial distribution. Both are used to model the number of successes, given the following:

- A fixed number of trials
- Two possible outcomes on each trial

The difference is that the binomial distribution requires the probability of success to be the same for all trials, whereas the hypergeometric distribution does not. Consider drawing from a deck of cards. If 5 cards are drawn, the probability of getting exactly 2 hearts can be computed using the binomial distribution if after each draw the card is replaced in the deck and the cards are shuffled. By replacing the cards and shuffling, the probability of getting a heart on each of the 5 draws remains fixed at $\frac{13}{52}$. If each card is not replaced after it is drawn, the probability of getting a heart on the first draw is $\frac{13}{52}$, but the probability of getting a heart on the second draw is dependent on the outcome of the first draw. If the first draw resulted in a heart, the probability of getting a heart on the second draw is 12/51. If the first draw did not result in a heart, the probability of getting a heart on the second draw is 13/51. The hypergeometric distribution is used to model this situation. This is also why the hypergeometric distribution is referred to as the distribution that models sampling without replacement.

The hypergeometric probability density function is

$$p(x, N, n, m) = \frac{\binom{m}{x}\binom{N-m}{n-x}}{\binom{N}{n}} \tag{3.31}$$

where $p(x, N, n, m)$ is the probability of exactly x successes in a sample of n drawn from a population of N containing m successes. The hypergeometric cumulative distribution function is

$$P(x, N, n, m) = \sum_{i=0}^{x} \frac{\binom{m}{i}\binom{N-m}{n-i}}{\binom{N}{n}} \tag{3.32}$$

The mean and the variance of the hypergeometric distribution are, respectively,

$$\mu = \frac{nm}{N} \tag{3.33}$$

$$\sigma^2 = \left(\frac{nm}{N}\right)\left(1 - \frac{m}{N}\right)\left(\frac{N-n}{N-1}\right) \tag{3.34}$$

Example 3.14: Fifty items are submitted for acceptance. If it is known that there are 4 defective items in the lot, what is the probability of finding exactly 1 defective item in a sample of 5? What is the probability of finding fewer than 2 defective items in a sample of 5?

Solution: For this example,

$$x = 1$$

$$N = 50$$

$$n = 5$$

$$m = 4$$

The probability of finding exactly 1 defective item in a sample of 5 is

$$p(1, 50, 5, 4) = \frac{\binom{4}{1}\binom{50-4}{5-1}}{\binom{50}{5}} = \frac{(4)(163{,}185)}{2{,}118{,}760} = 0.30808$$

This answer can be found using the Microsoft Excel function

=HYPGEOMDIST(1,5,4,50)

The probability of finding fewer than 2 defective items in a sample of 5 is equal to the probability of exactly 0 plus the probability of exactly 1. The probability of exactly 0 is

$$p(0, 50, 5, 4) = \frac{\binom{4}{0}\binom{50-4}{5-0}}{\binom{50}{5}} = \frac{(1)(1{,}370{,}754)}{2{,}118{,}760} = 0.64696$$

The probability of finding fewer than 2 defective items in a sample of 5 is

$$0.30808 + 0.64696 = 0.95504$$

Unfortunately, Microsoft Excel does not have a cumulative form of the hypergeometric distribution.

The binomial distribution can be used to approximate the hypergeometric distribution when the population is large with respect to the sample size. When N is much larger than n, the change in the probability of success on a single trial is too small to significantly affect the results of the calculations. Again, it is silly to use approximations when exact solutions can be found with electronic spreadsheets. These approximations were useful for the engineer toiling with a slide rule but are of little use now.

Geometric Distribution

The geometric distribution is similar to the binomial distribution, in that the probability of occurrence is constant from trial to trial and the trials are independent. The binomial distribution models situations where the number of trials is fixed, and the random variable is the number of successes. The geometric distribution requires exactly 1 success, and the random variable is the number of trials required to obtain the first success. The geometric distribution is a special case of the negative binomial distribution. The negative binomial distribution models the number of trials required to obtain m successes, and m is not required to be equal to 1.

The geometric probability density function is

$$p(x, p) = p(1 - p)^{(x-1)} \tag{3.35}$$

where $p(x, p)$ is the probability that the first success occurs on the xth trial given a probability of success on a single trial of p. The probability that more than n trials are required to obtain the first success is

$$p(x > n) = (1 - p)^n \tag{3.36}$$

The mean and variance of the geometric distribution are, respectively,

$$\mu = \frac{1}{p} \tag{3.37}$$

$$\sigma^2 = \frac{1 - p}{p^2} \tag{3.38}$$

Example 3.15: The probability of an enemy aircraft penetrating friendly airspace is 0.01. What is the probability that the first penetration of friendly airspace is accomplished by the 80th aircraft to attempt the penetration of friendly airspace?

Solution: The probability that the first success occurs on the 80th trial with a probability of success of 0.01 on each trial is

$$p(80, 0.01) = 0.01(1 - 0.01)^{(80-1)} = 0.0452$$

Example 3.16: The probability of an enemy aircraft penetrating friendly airspace is 0.01. What is the probability that it will take more than 80 attempts to penetrate friendly airspace?

Solution: The probability that it will take more than 80 attempts to penetrate friendly airspace with a probability of success of 0.01 on each trial is

$$p(x > 80) = (1 - 0.01)^{80} = 0.4475$$

Identifying the Correct Discrete Distribution

It is often difficult to determine the correct discrete distribution when faced with a statistical problem. To make this task easier, ask the following questions.

1. Is a rate being modeled, such as defects per car or rabbits per acre, and is there no upper bound on the number of possible occurrences? (For example, is it conceivable that there could be 10,000 defects on a car? There is no upper limit imposed by a specific number of trials. If there are 5 trials, the upper limit on the number of successes is 5.) If the answer is yes, the Poisson distribution is probably the appropriate distribution. If the answer is no, go to question 2.

2. Is there a fixed number of trials? If yes, go to question 3. If no, is there a fixed number of successes, with the number of trials being the random variable? If the answer is yes, then if the number of successes is fixed at 1, use the geometric distribution; if the number of successes is fixed at a value greater than 1, use the negative binomial distribution.

3. Is the probability of success the same on all trials? If yes, use the binomial distribution; if no, use the hypergeometric distribution.

The flowchart shown in Figure 3.16 summarizes these questions.

Sampling Distributions

The chi-square, *t*-, and *F*-distributions are formed from combinations of random variables. Because of this, they generally are not used to model physical phenomena, such as time to fail, but are used to make decisions and to construct confidence intervals.

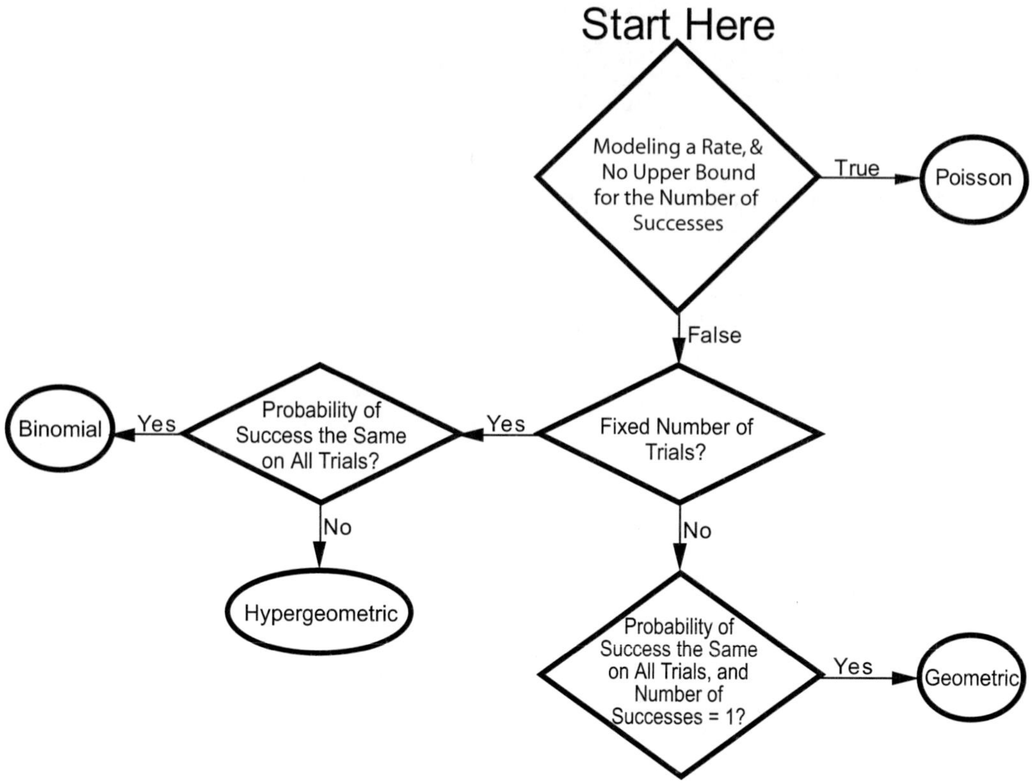

Figure 3.16 *Choosing a discrete distribution.*

Chi-Square Distribution

The chi-square distribution is formed by summing the square of standard normal random variables. For example, if z is a standard normal random variable, then

$$y = z_1^2 + z_2^2 + z_3^2 + \ldots + z_n^2 \tag{3.39}$$

is a chi-square random variable with n degrees of freedom.

A chi-square random variable is also created by summing two or more chi-square random variables. A distribution having this property is regenerative. The chi-square distribution is a special case of the gamma distribution with a failure rate of 2, and degrees of freedom equal to 2 divided by the number of degrees of freedom for the corresponding chi-square distribution.

The chi-square probability density function is

$$f(x) = \frac{x^{\left(\frac{\nu}{2}-1\right)} e^{-\frac{x}{2}}}{2^{\frac{\nu}{2}} \Gamma\left(\frac{\nu}{2}\right)}, \quad x > 0 \tag{3.40}$$

where ν is the degrees of freedom, and $\Gamma(x)$ is the gamma function. Figure 3.17 shows the chi-square probability density function.

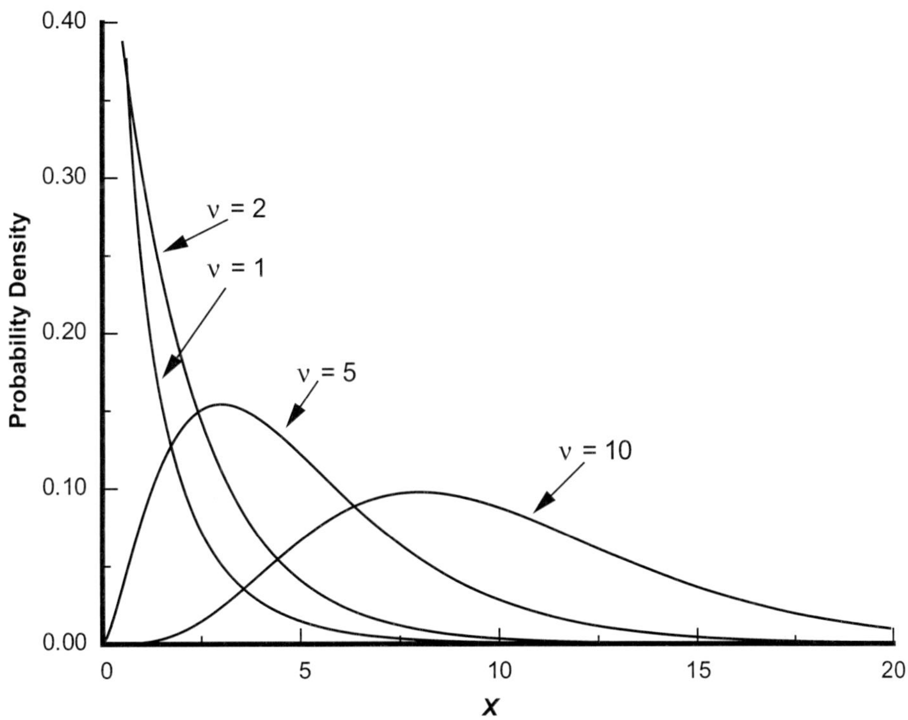

Figure 3.17 *Chi-square probability density function.*

The critical value of the chi-square distribution is given in Appendix A, but it is easier to use the Microsoft Excel function

=CHIINV(probability,degrees_freedom)

Example 3.17: A chi-square random variable has 7 degrees of freedom. What is the critical value if 5% of the area under the chi-square probability density is desired in the right tail?

Solution: When hypothesis testing, this is commonly referred to as the critical value with 5% significance, or $\alpha = 0.05$. From Appendix A, this value is 14.067. This value is also found using the Microsoft Excel function

$$=\text{CHIINV}(0.05,7)$$

t-*Distribution*

The t-distribution is formed by combining a standard normal random variable and a chi-square random variable. If z is a standard normal random variable, and χ^2 is a chi-square random variable with ν degrees of freedom, then a random variable with a t-distribution is

$$t = \frac{z}{\sqrt{\frac{\chi^2}{\nu}}} \tag{3.41}$$

The t-distribution is equivalent to the F-distribution with 1 and ν degrees of freedom. The t-distribution is commonly used for hypothesis testing and constructing confidence intervals for means. It is used in place of the normal distribution when the standard deviation is unknown. The t-distribution compensates for the error in the estimated standard deviation. If the sample size is large, $n > 100$, the error in the estimated standard deviation is small, and the t-distribution is approximately normal. The t-probability density function is

$$f(x) = \frac{\Gamma\left[\frac{(\nu+1)}{2}\right]}{\Gamma\left(\frac{\nu}{2}\right)\sqrt{\Pi\nu}}\left(1 + \frac{x^2}{\nu}\right)^{-\frac{(\nu+1)}{2}}, \quad -\infty < x \tag{3.42}$$

where ν is the degrees of freedom. Figure 3.18 shows the t-probability density function.

The mean and variance of the t-distribution are, respectively,

$$\mu = 0 \tag{3.43}$$

$$\sigma^2 = \frac{\nu}{\nu - 2}, \quad \nu \ge 3 \tag{3.44}$$

From a random sample of n items, the probability that

$$t = \frac{\bar{x} - \mu}{\frac{s}{\sqrt{n}}} \tag{3.45}$$

falls between any two specified values is equal to the area under the t-probability density function between the corresponding values on the x axis with $n - 1$ degrees of freedom.

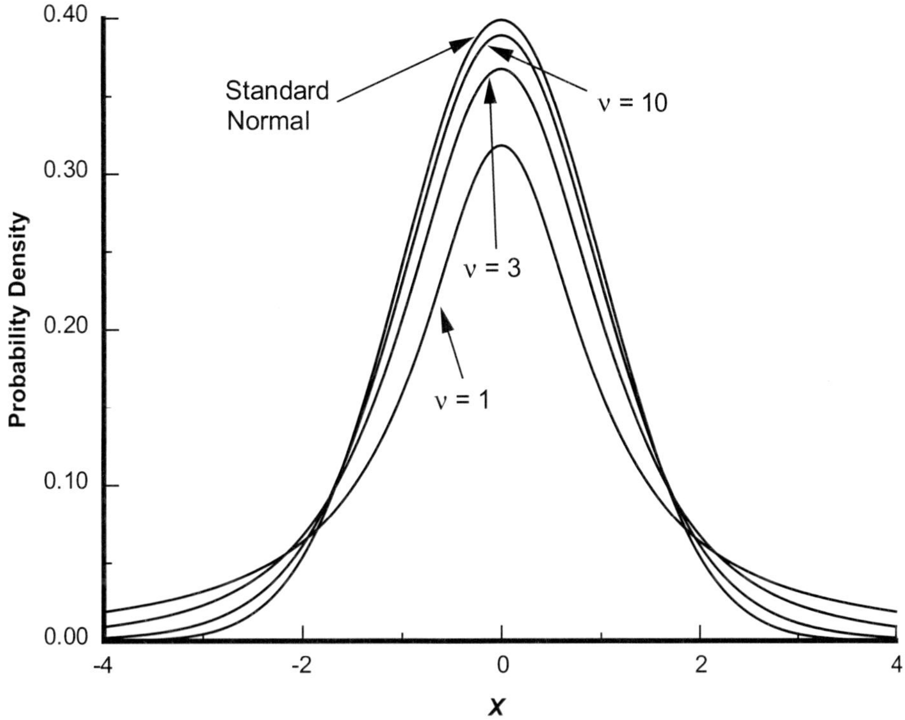

Figure 3.18 *The* t-*probability density function.*

Example 3.18: The burst strength of 15 randomly selected seals is given below. What is the probability that the average burst strength of the population is greater than 500?

480	489	491	508	501
500	486	499	479	496
499	504	501	496	498

Solution: The mean of these 15 data points is 495.13. The sample standard deviation of these 15 data points is 8.467. The probability that the population mean is greater than 500 is equal to the area under the *t*-probability density function, with 14 degrees of freedom, to the left of

$$t = \frac{495.13 - 500}{\frac{8.467}{\sqrt{15}}} = -2.227$$

From Appendix A, the area under the t-probability density function, with 14 degrees of freedom, to the left of –2.227, is 0.0215. This value must be interpolated using Appendix A but can be computed directly in Microsoft Excel using the function

$$=\text{TDIST}\left(-2.227,14,1\right)$$

Simply stated, making an inference from the sample of 15 data points, there is a 2.15% that the true population mean is greater than 500.

F-*Distribution*

If X is a chi-square random variable with ν_1 degrees of freedom, Y is a chi-square random variable with ν_2 degrees of freedom, and X and Y are independent, then

$$F = \frac{\frac{X}{\nu_1}}{\frac{Y}{\nu_2}} \tag{3.46}$$

is F-distributed with ν_1 and ν_2 degrees of freedom.

The F-distribution is used extensively to test for equality of variances from two normal populations. If U and V are the variances of independent random samples of size n and m taken from normally distributed populations with variances of w and z, then

$$F = \frac{\frac{U}{w}}{\frac{V}{z}} \tag{3.47}$$

is a random variable with an F-distribution with $\nu_1 = n - 1$ and $\nu_2 = m - 1$. The F-probability density function is

$$f(x) = \left(\frac{\Gamma\left(\frac{\nu_1+\nu_2}{2}\right)\left(\frac{\nu_1}{\nu_2}\right)^{\frac{\nu_1}{2}}}{\Gamma\left(\frac{\nu_1}{2}\right)\Gamma\left(\frac{\nu_2}{2}\right)}\right)\left(\frac{x^{\left(\frac{\nu_1}{2}-1\right)}}{\left(1+\frac{\nu_1 x}{\nu_2}\right)^{\frac{(\nu_1+\nu_2)}{2}}}\right), \quad x > 0 \tag{3.48}$$

Figure 3.19 shows the F-probability density function.

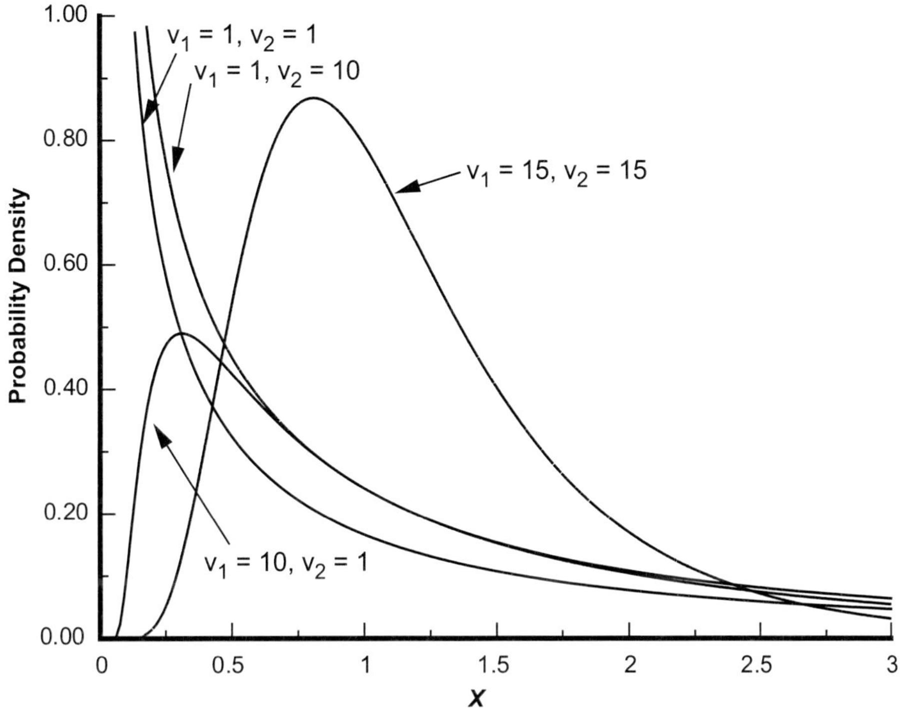

Figure 3.19 *The* F-*probability density function.*

The F-cumulative distribution function is given in Appendix A. Only the upper tail is given, but the other tail can be computed using the expression

$$F_{\alpha,v_1,v_2} = \frac{1}{F_{1-\alpha,v_2,v_1}} \tag{3.49}$$

It is easier to find critical values of the F-distribution using the Microsoft Excel function

$$=\text{FINV}\left(\alpha, v_1, v_2\right)$$

Example 3.19: Find $F_{0.05}$ with $v_1 = 9$ and $v_2 = 10$, and find $F_{0.95}$ with $v_1 = 10$ and $v_2 = 9$.

Solution: From Appendix A, $F_{0.05}$ with $v_1 = 9$ and $v_2 = 10$ is 3.02. $F_{0.95}$ with $v_1 = 10$ and $v_2 = 9$ is equal to the inverse of $F_{0.05}$ with $v_1 = 9$ and $v_2 = 10$. Thus,

$$F_{0.95,10,9} = \frac{1}{F_{0.05,9,10}} = \frac{1}{3.02} = 0.331$$

The following Microsoft Excel functions provide the solutions, respectively,

$$=\text{FINV}(0.05,9,10)$$

$$=\text{FINV}(0.95,10,9)$$

Bayes' Theorem

Previous discussions of probability distributions in this chapter assumed no prior knowledge of events. The event being modeled, such as time to fail, was not influenced by the knowledge of some other event. Bayes' theorem is based on a prior distribution and a posterior distribution.

Consider the life of a human being. If a person is selected at random, what is the probability that this person will live to be 80 years old? Now if it is known that this randomly selected person has heart disease, the probability that this person survives to be 80 years old changes. With no knowledge of the person, estimates of life expectancy are based on the prior distribution. After the influence of the knowledge that the person has heart disease is considered, estimates are made using the posterior distribution.

For example, an item has a normally distributed time-to-fail distribution with a mean of 800 hours and a standard deviation of 200. What is the probability of an item surviving for 900 hours? Given an item has survived for 850 hours, what is the probability of surviving for 900 hours? The first question is answered using the prior distribution, and the second question is answered using the posterior distribution. Figure 3.20 shows the prior and posterior probability density functions.

Figure 3.21 shows the prior and posterior reliability functions.

Given a prior probability density function of $f(x)$ and survival until x_0, the posterior probability density function is

$$f_{\text{posterior}}(x) = \frac{f_{\text{prior}}(x)}{R(x_0)} \tag{3.50}$$

For the previous example, the probability of survival for 850 hours is equal to the area under the standard normal probability density function to the right of $z = 0.25$,

$$z = \frac{x-\mu}{\sigma} = \frac{850-800}{200} = 0.25$$

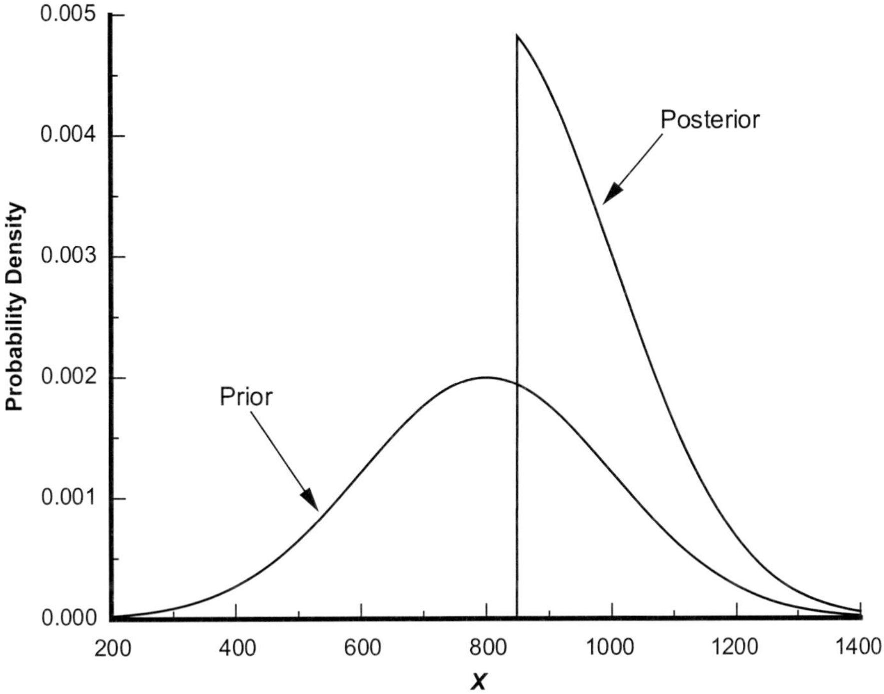

Figure 3.20 *Prior and posterior probability density functions.*

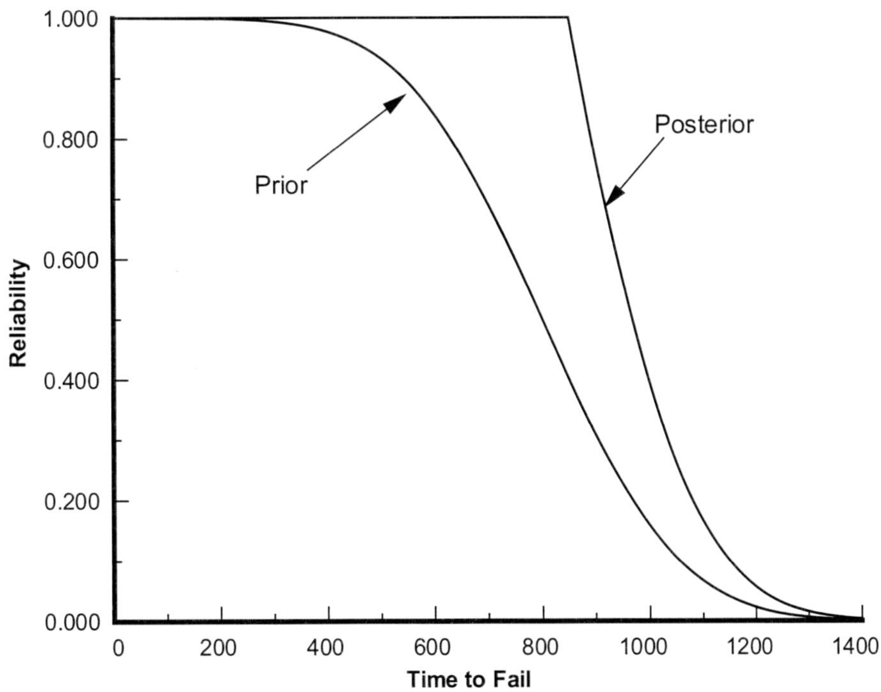

Figure 3.21 *Prior and posterior reliability functions.*

From Appendix A, the area under the standard normal probability density function to the right of $z = 0.25$ is 0.4013.* Thus, given survival for 850 hours, the posterior probability density function for the previous example is

$$f_{\text{posterior}}(x) = \frac{\phi(x)}{0.4013}$$

where $\phi(x)$ is the standard normal probability density function.

Given a prior reliability function of $f(x)$ and survival until x_0, the posterior reliability function is

$$R_{\text{posterior}}(x) = \frac{R_{\text{prior}}(x)}{R(x_0)} \tag{3.51}$$

Stated another way, if an item has survived until x_0, the probability of survival until x_1 is

$$R(x_1 \backslash x_0) = \frac{R(x_1)}{R(x_0)} \tag{3.52}$$

Continuing with the example, the probability of survival for 850 hours is 0.4013. The probability of survival for 900 hours given survival for 850 hours is equal to the reliability at 850 hours divided by the reliability at 900 hours. The probability of survival for 900 hours is equal to the area under the standard normal probability density function to the right of $z = 0.5$,

$$z = \frac{x - \mu}{\sigma} = \frac{900 - 800}{200} = 0.5$$

From Appendix A, the area under the standard normal probability density function to the right of $z = 0.5$ is 0.3085. Thus, given survival for 850 hours, the probability of survival for 900 hours is

$$R(900 \backslash 800) = \frac{R(900)}{R(850)} = \frac{0.3085}{0.4013} = 0.76$$

Example 3.20: A resistor has an exponential time-to-fail distribution with a failure rate of 0.02 hours. A resistor has survived for 100 hours. What is the probability of survival for 200 hours?

Solution: The probability of survival for 200 hours, given survival for 100 hours, is

$$R(200 \backslash 100) = \frac{R(200)}{R(100)} = \frac{e^{-(0.02)(200)}}{e^{-(0.02)(100)}} = \frac{0.0183156}{0.135335} = 0.1353$$

* This value can be found using the Microsoft Excel function =-NORMDIST$(850,800,200,1)$.

Note that the probability of survival in the interval from 0 to 100 hours is equal to the probability of survival in the interval from 100 to 200 hours, given survival for 100 hours. This is known as the "lack of memory" property, which is unique to the exponential and Poisson distributions.

A more general form of Bayes' theorem is

$$P(A_i \backslash B) = \frac{P(A_i)P(B \backslash A_i)}{\sum_{i=1}^{n} P(A_i)P(B \backslash A_i)} \tag{3.53}$$

Example 3.21: Three jars each contain 3 coins. Jar 1 contains 3 quarters; Jar 2 contains 1 quarter, 1 nickel, and 1 penny; and Jar 3 contains 3 pennies. If a coin is drawn from a randomly selected jar, what is the probability the coin was drawn from Jar 3 if the selected coin was a penny?

Solution: The probability of selecting any of the 3 jars with no knowledge of the coin drawn is $\frac{1}{3}$. Given the coin was selected from Jar 1, the probability of the coin being a penny is 0; there are no pennies in Jar 1. Given the coin was selected from Jar 2, the probability of the coin being a penny is $\frac{1}{3}$; there are 3 coins in Jar 2, and 1 of the 3 coins is a penny. Given the coin was selected from Jar 3, the probability of the coin being a penny is 1; all 3 coins in Jar 3 are pennies. The probability that the coin was drawn from Jar 3, given the coin was a penny, is

$$P(\text{Jar 3}\backslash\text{a penny was drawn}) = \frac{\left(\frac{1}{3}\right)(1)}{\left(\frac{1}{3}\right)(0) + \left(\frac{1}{3}\right)\left(\frac{1}{3}\right) + \left(\frac{1}{3}\right)(1)} = \frac{3}{4}$$

Summary

Distributions can be used to model (Weibull, exponential, normal, lognormal, Poisson, binomial, hypergeometric, geometric) or for statistical inference (normal, *t*, chi-square, *F*).

It is important to verify that the correct distribution is being used. Improper distribution choice will probably result in errors. The errors may be small or large, depending on the situation. Understand the statistical distributions, and verify the distribution fit and any assumptions before proceeding.

CHAPTER 4

PARAMETER ESTIMATION

Many methods are available for parameter estimation. In reliability engineering, the most popular methods are the following:

- Maximum likelihood estimation
- Probability plotting
- Hazard plotting

It is desirable for a parameter estimator to have the following properties:

1. **Lack of Bias**—If the expected value of the estimator is equal to the true value of the parameter, it is said to be unbiased.

2. **Minimum Variance**—The smaller the variance of the estimate, the smaller the sample size required to obtain the level of accuracy desired, and the more efficient the estimator. The most efficient estimator is the estimator with minimum variance.

3. **Consistency**—As the sample size is increased, the value of the estimated parameter becomes closer to the true value of the parameter.

4. **Sufficiency**—The estimator uses all information available in the data set.

The method used to estimate parameters depends on the type of data or testing involved and the distribution of interest. In addition to these parameter estimation methods, this chapter describes censored data and presents methods of parameter estimation for the exponential, normal, lognormal, and Weibull distributions. Methods are presented for complete and censored data.

Maximum Likelihood Estimation

Maximum likelihood is the most widely used method of generating estimators. It is based on the principle of determining the parameter(s) value(s) that maximize(s) the probability of obtaining the sample data.

The likelihood function for a given distribution is a representation of the probability of obtaining the sample data. Let $x_1, x_2, \dots, x_n$ be independent random variables from the probability density function $f(x,\theta)$, where θ is the single distribution parameter. Then

$$L(x_1, x_2, \dots, x_n; \theta) = f(x_1,\theta)\, f(x_2,\theta) \dots f(x_n,\theta) \tag{4.1}$$

is the joint distribution of the random variables, or the likelihood function. The maximum likelihood estimate, $\hat{\theta}$, maximizes the likelihood function. This estimate is asymptotically normal. Often, the natural logarithm of the likelihood function is maximized to simplify computations.

The variances of the estimates can be found by inverting the matrix of the negative of the second partial derivatives of the likelihood function, also known as the local information matrix. These estimates are asymptotically normal, and the variances obtained from the local information matrix are used to calculate confidence intervals.

Probability Plotting

Probability plotting is a graphical method of parameter estimation. For the assumed distribution, the cumulative distribution function is transformed to a linear expression, usually by a logarithmic transformation, and plotted. If the plotted points form a straight line, the assumed distribution is acceptable, and the slope and the intercept of the plot provide the information needed to estimate the parameters of the distribution of interest. The median rank is usually used to estimate the cumulative distribution function, although there are several alternatives such as the mean rank and the Kaplan-Meier product limit estimator.

If manually constructing a probability plot, distribution-specific probability paper is required. By using probability paper, the failure times and cumulative distribution function estimates can be plotted directly. With the power of personal computers and electronic spreadsheets, specialized graph paper is no longer needed because the necessary transformations can be made quickly and easily.

Hazard Plotting

Hazard plotting is a graphical method of parameter estimation. The cumulative hazard function is transformed to a linear expression, usually by a logarithmic transformation, and plotted. The slope and the intercept of the plot provide the information needed to estimate the parameters of the distribution of interest.

If manually constructing a hazard plot, distribution-specific hazard paper is required. By using hazard paper, the failure times and cumulative hazard function estimates can be plotted directly. With the power of personal computers and electronic spreadsheets, specialized graph paper is no longer needed because the necessary transformations can be made quickly and easily.

Exponential Distribution

The simplest method of parameter estimation for the exponential distribution is the method of maximum likelihood. Maximum likelihood provides an unbiased estimate but no indication of goodness of fit. Graphical methods, although more involved, provide a visual goodness-of-fit test. Often, graphical methods will be used in conjunction with maximum likelihood estimation.

Maximum Likelihood Estimation

The exponential probability density function is

$$f(x) = \frac{1}{\theta} e^{-\frac{x}{\theta}}, \quad x \geq 0 \tag{4.2}$$

The maximum likelihood estimation for the parameter θ is

$$\hat{\theta} = \frac{\sum_{i=1}^{n} x_i}{r} \tag{4.3}$$

where

x_i is the ith data point (this may be a failure or a censoring point)

n is the total number of data points (both censored and uncensored)

r is the number of failures

This estimate is unbiased and is the minimum variance estimator.

Example 4.1: The cycles to fail for seven springs are:

30,183	14,871	35,031	76,321
43,891	31,650	12,310	

Assuming an exponential time-to-fail distribution, estimate the mean time to fail and the mean failure rate.

Solution: The mean time to fail is

$$\hat{\theta} = \frac{30{,}183 + 14{,}871 + 35{,}031 + 76{,}321 + 43{,}891 + 31{,}650 + 12{,}310}{7}$$

$$= \frac{244{,}257}{7} = 34{,}893.9 \text{ cycles}$$

The mean failure rate is the inverse of the mean time to fail,

$$\hat{\lambda} = \frac{1}{34{,}893.9} = 0.0000287 \text{ failures per cycle}$$

Example 4.2: Assume the data in Example 4.1 represent cycles to fail for seven springs, but an additional 10 springs were tested for 80,000 cycles without failure. Estimate the mean time to fail and the mean failure rate.

Solution: The mean time to fail is

$$\hat{\theta} = \frac{244,257 + 10(80,000)}{7} = 149,179.6 \text{ cycles}$$

The mean failure rate is

$$\lambda = \frac{1}{149,179.6} = 0.0000067 \text{ failures per cycle}$$

For a time truncated test, a confidence interval for θ is

$$\frac{2\sum_{i=1}^{n} x_i}{\chi^2_{\left(\frac{\alpha}{2},2r+2\right)}} \le \theta \le \frac{2\sum_{i=1}^{n} x_i}{\chi^2_{\left(1-\frac{\alpha}{2},2r\right)}} \tag{4.4}$$

Note that the χ^2 degrees of freedom differ for the upper and lower limits.

Example 4.3: Fifteen items were tested for 1,000 hours. Failures occurred at times of 120 hours, 190 hours, 560 hours, and 812 hours. Construct a 90% confidence interval for the mean time to fail and the failure rate.

Solution: This is a time truncated test. The mean life estimate is

$$\hat{\theta} = \frac{120 + 190 + 560 + 812 + 11(1,000)}{4} = \frac{12,682}{4} = 3,170.5$$

For a 90% confidence interval,

$$\alpha = 0.1$$

$$\chi^2_{(0.05,10)} = 18.307$$

$$\chi^2_{(0.95,8)} = 2.733$$

The 90% confidence interval for θ is

$$\frac{2(12,682)}{18.307} \le \theta \le \frac{2(12,682)}{2.733}$$

$$1,385.5 \le \theta \le 9,280.6$$

The confidence interval for the failure rate is the inverse of the confidence interval for the mean time to fail,

$$\frac{1}{9,280.6} \leq \lambda \leq \frac{1}{1,385.5}$$

$$0.0001077 \leq \lambda \leq 0.0007217$$

For a failure truncated test and for multiple censored data, a confidence interval for θ is

$$\frac{2\sum_{i=1}^{n} x_i}{\chi^2_{\left(\frac{\alpha}{2},2r\right)}} \leq \theta \leq \frac{2\sum_{i=1}^{n} x_i}{\chi^2_{\left(1-\frac{\alpha}{2},2r\right)}} \tag{4.5}$$

Note that the χ^2 degrees of freedom are the same for the upper and lower limits.

Example 4.4: Twelve items were tested with failures occurring at times of 43 hours, 67 hours, 92 hours, 94 hours, and 149 hours. At a time of 149 hours, the testing was stopped for the remaining seven items. Construct a 95% confidence interval for the mean time to fail.

Solution: This is a failure truncated test. The mean life estimate is

$$\hat{\theta} = \frac{43 + 67 + 92 + 94 + 149 + 7(149)}{5} = \frac{1,488}{5} = 297.6$$

For a 95% confidence interval,

$$\alpha = 0.05$$

$$\chi^2_{(0.025,10)} = 20.483$$

$$\chi^2_{(0.975,10)} = 3.247$$

The 95% confidence interval for θ is

$$\frac{2(1,488)}{20.483} \leq \theta \leq \frac{2(1,488)}{3.247}$$

$$145.3 \leq \theta \leq 916.5$$

For failure-free testing, the one-sided lower confidence limit simplifies to

$$\frac{-nt}{\ln \alpha} \leq \theta \tag{4.6}$$

where

t is the testing time

ln is the natural logarithm

α is the significance ($\alpha = 0.05$ for a 95% limit)

Example 4.5: Twenty items are tested for 230 hours without failure. Determine a 90% lower confidence limit for θ.

Solution:

$$\frac{-20(230)}{\ln(0.1)} = 1{,}997.8 \text{ hours}$$

A confidence interval for reliability is

$$e^{-\frac{x}{\theta_L}} \leq R(x) \leq e^{-\frac{x}{\theta_U}} \tag{4.7}$$

where

θ_L is the lower confidence limit for the mean time to fail

θ_U is the upper confidence limit for the mean time to fail

A confidence interval for percentiles is

$$-\theta_L \ln(1-P) \leq x \leq -\theta_U \ln(1-P) \tag{4.8}$$

where P is the probability of failure prior to time = x.

Example 4.6: Twenty items are tested for 230 hours without failure. Determine a 90% lower confidence limit for reliability at time = 1,000.

Solution: The lower 90% confidence limit for the mean is

$$\frac{-20(230)}{\ln(0.1)} = 1{,}997.8 \text{ hours}$$

The lower 90% confidence limit for reliability at time = 1,000 is

$$R_L(1{,}000) = e^{-\frac{1{,}000}{1{,}997.8}} = 0.606$$

Hazard Plotting

The exponential cumulative hazard function is

$$H(x) = \frac{x}{\theta} \tag{4.9}$$

If a data set is exponentially distributed, a plot of the exponential cumulative hazard function yields a linear fit with a zero intercept and a slope of $\frac{1}{\theta}$. To construct a hazard plot, an estimate for $H(x)$ is needed. The cumulative hazard function is estimated by the cumulative of the inverse of the reverse ranks. For a data set of n points ordered from smallest to largest, the first point has a rank of n, the second $n-1$, and so forth.

Example 4.7: Construct a hazard plot for the following failure data, given that an additional 7 items were tested for 149 cycles without failure:

43, 67, 92, 94, 149

Solution: Table 4.1 shows the calculations for the values necessary for the hazard plot. Figure 4.1 shows the plot. Note that censored data points are not plotted. The slope of the best-fit straight line through the data with an intercept of 0 is 0.00333. The estimated mean is

$$\theta = \frac{1}{0.00333} = 300.3$$

TABLE 4.1
TABULATIONS FOR AN EXPONENTIAL HAZARD PLOT

Time to Fail	Reverse Rank	$h(t)$	$H(t)$
43	12	0.0833	0.0833
67	11	0.0909	0.1742
92	10	0.1000	0.2742
94	9	0.1111	0.3854
149	8	0.1250	0.5104
149 c	7		
149 c	6		
149 c	5		
149 c	4		
149 c	3		
149 c	2		
149 c	1		

c = Censored.

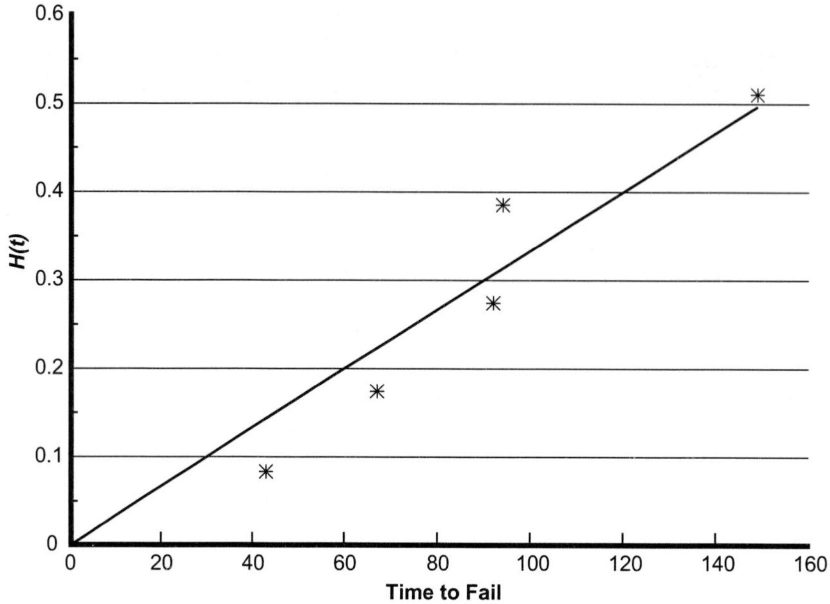

Figure 4.1 *Hazard plot for exponential data.*

Probability Plotting

The exponential cumulative distribution function is

$$F(x) = 1 - e^{-\frac{x}{\theta}} \tag{4.10}$$

By manipulating this expression algebraically, it can be transformed to a linear format,

$$\ln \frac{1}{1 - F(x)} = \frac{x}{\theta} \tag{4.11}$$

If a data set follows an exponential distribution, a plot of $\ln \frac{1}{1 - F(x)}$ versus x will be linear with a 0 intercept and a slope of $\frac{1}{\theta}$. Before a plot can be constructed, an estimate for $F(x)$ is needed. The cumulative distribution function, $F(x)$, is usually estimated from the median rank, but other estimates such as the mean rank and the Kaplan-Meier product limit estimator are also used. The median rank estimate for $F(x)$ is

$$\hat{F}(x) = \frac{O_i - 0.3}{n + 0.4} \tag{4.12}$$

where O_i is the modified order of failure of the ith data point.

A modified order of failure is needed only if censored data are involved; if not, the original order of failure, *i*, is equivalent to the modified order of failure. The logic for a modified order of failure is as follows. Consider three items: the first was tested for 3 hours, and the test was stopped without failure; the second item was tested and failed after 4 hours; and the third item was tested and failed after 4.5 hours. For this data set, the failure order is unclear. The first item could have been either the first failure, the second failure, or the third failure. Thus, it is not certain that the first item to fail, the second item, is the first ordered failure. The modified order of failure is computed from the expression

$$O_i = O_{i-1} + I_i \tag{4.13}$$

where I_i is the increment for the ith failure and is computed from the expression

$$I_i = \frac{(n+1) - O_p}{1+c} \tag{4.14}$$

where

n is the total number of points in the data set (both censored and uncensored)

c is the number of points remaining in the data set (including the current point)

O_p is the order of the previous failure

An alternative to plotting x versus $\ln\frac{1}{1-F(x)}$ on conventional graph paper is to plot x versus $F(x)$ on specialized probability paper. The advantage of probability paper is that the values of $\ln\frac{1}{1-F(x)}$ do not have to be computed. With computers, this technique is obsolete.

Example 4.8: Construct a probability plot for the following failure data, given that an additional 7 items were tested for 149 cycles without failure:

43, 67, 92, 94, 149

Solution: Table 4.2 contains the calculations necessary for plotting. Figure 4.2 shows the probability plot. The slope of the best-fit straight line through the origin is 0.00304, which estimates the failure rate for the exponential distribution. The mean of the distribution is

$$\theta = \frac{1}{0.00304} = 328.9$$

Often, reliability confidence limits are added to probability plots. Upper and lower confidence limits are approximated by 5% and 95% ranks. These ranks can be taken from the expression

$$w_\alpha = \frac{\dfrac{j}{n-j+1}}{F_{1-\alpha,2(n-j+1),2j} + \dfrac{j}{n-j+1}} \tag{4.15}$$

TABLE 4.2
TABULATIONS FOR EXPONENTIAL PROBABILITY PLOTTING.

Time to Fail	O_i	Median Rank, $F(t)$	$\frac{1}{1-F(t)}$	$\ln\frac{1}{1-F(t)}$
43	1	0.0565	1.0598	0.0581
67	2	0.1371	1.1589	0.1475
92	3	0.2177	1.2784	0.2456
94	4	0.2984	1.4253	0.3544
149	5	0.3790	1.6104	0.4765
149 c				
149 c				
149 c				
149 c				
149 c				
149 c				
149 c				

c = Censored.

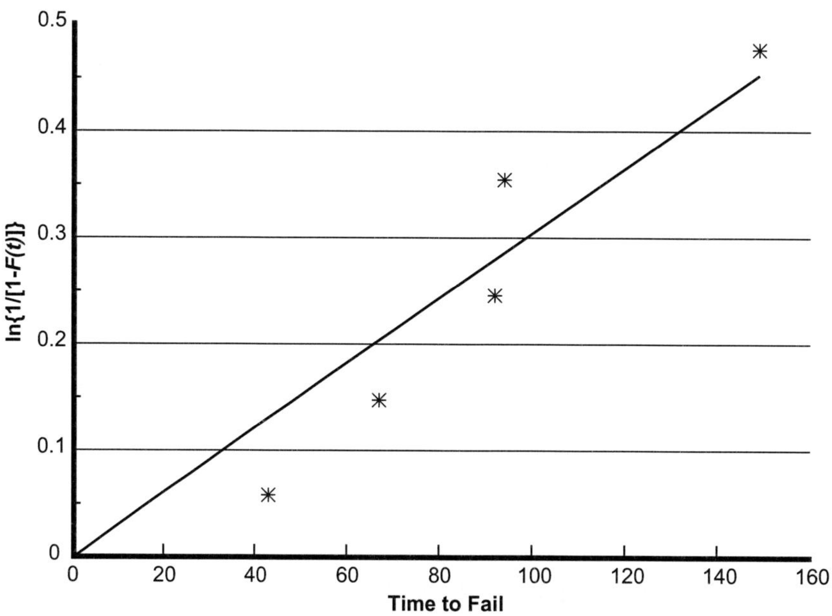

Figure 4.2 *Exponential probability plot.*

where

w_α is the $100(1-\alpha)\%$ nonparametric confidence limit

j is the failure order

n is the total number of data points (both censored and uncensored)

F_{α,ν_1,ν_2} is the critical value from the F-distribution

When multiple censored data are encountered, the modified failure orders will not be integers, and the rank values will have to be interpolated. The rank values are not plotted against the corresponding failure time. Any deviation of the failure time from the best-fit straight line through the data is considered sampling error, and the time against which the rank values are plotted is found by moving parallel to the x-axis until the best-fit straight line is intersected. This plotting position is

$$x_i = \theta \ln\left(\frac{1}{1-F(x_i)}\right) \tag{4.16}$$

Normal Distribution

The normal probability density function is

$$f(x) = \frac{1}{\sigma\sqrt{2\pi}}\exp\left[-\frac{1}{2}\left(\frac{x-\mu}{\sigma}\right)^2\right], \quad -\infty < x < \infty \tag{4.17}$$

where

μ is the distribution mean

σ is the distribution standard deviation

If no censoring is involved, the distribution mean is estimated from the expression

$$\hat{\mu} = \bar{x} = \frac{\sum_{i=1}^{n} x_i}{n} \tag{4.18}$$

where n is the sample size.

If no censoring is involved, the distribution standard deviation is estimated from the expression

$$\hat{\sigma} = \sqrt{\frac{n\sum_{i=1}^{n} x_i^2 - \left(\sum_{i=1}^{n} x_i\right)^2}{n(n-1)}} \tag{4.19}$$

However, when censored data are involved, parameter estimation becomes complicated. Three popular methods for parameter estimation for the normal distribution when censored data are encountered are as follows:

1. Maximum likelihood estimation
2. Hazard plotting
3. Probability plotting

The following sections present each of these alternatives.

Maximum Likelihood Estimation

The maximum likelihood equations for the normal distribution are

$$\frac{\partial L}{\partial \mu} = \frac{r}{\sigma}\left[\frac{\bar{x} - \mu}{\sigma} + \sum_{i=1}^{k} \frac{h(x_i)}{r}\right] = 0 \tag{4.20}$$

$$\frac{\partial L}{\partial \sigma} = \frac{r}{\sigma}\left[\frac{s^2 + (\bar{x} - \mu)^2}{\sigma^2} - 1 + \sum_{i=1}^{k} \frac{z(x_i)h(x_i)}{r}\right] = 0 \tag{4.21}$$

where

r is the number of failures

k is the number of censored observations

$\bar{x}$ is the sample mean of the failures

s is the sample standard deviation for the failures

$z(x_i)$ is the standard normal deviate

$$z(x_i) = \frac{x_i - \mu}{\sigma}$$

$h(x_i)$ is the hazard function evaluated at the ith point

$$h(x_i) = \frac{\phi(z(x_i))}{\sigma\left[1 - \Phi(z(x_i))\right]}$$

where

> $\phi(z(x_i))$ is the standard normal probability density function evaluated at the ith point
>
> $\Phi(z(x_i))$ is the standard normal cumulative distribution function evaluated at the ith point

Note that if no censored data are involved, these expressions reduce to the sample mean and the sample standard deviation.

Iterative techniques are necessary to these equations. A standard method based on Taylor series expansions involves repeatedly estimating the parameters until a desired level of accuracy is reached. Estimates of μ and σ are given by the expressions, respectively,

$$\hat{\mu}_i = \hat{\mu}_{i-1} + h \tag{4.22}$$

$$\hat{\sigma}_i = \hat{\sigma}_{i-1} + k \tag{4.23}$$

where

> h is a correction factor for the distribution mean
>
> k is a correction factor for the distribution standard deviation

For each iteration, the correction factors are estimated from the expressions

$$h\frac{\partial^2 L}{\partial \mu^2} + k\frac{\partial^2 L}{\partial \mu \partial \sigma} = \frac{\partial L}{\partial \mu} \tag{4.24}$$

$$h\frac{\partial^2 L}{\partial \mu \partial \sigma} + k\frac{\partial^2 L}{\partial \sigma^2} = \frac{\partial L}{\partial \sigma} \tag{4.25}$$

where

$$\frac{\partial^2 L}{\partial \mu^2} = -\frac{r}{\sigma^2}\left[1 + \sum_{i=1}^{k}\frac{A_i}{r}\right] \tag{4.26}$$

$$\frac{\partial^2 L}{\partial \mu \partial \sigma} = -\frac{r}{\sigma^2}\left[\frac{2(\bar{x} - \mu)}{\sigma} + \sum_{i=1}^{k}\frac{B_i}{r}\right] \tag{4.27}$$

$$\frac{\partial^2 L}{\partial \sigma^2} = -\frac{r}{\sigma^2}\left[\frac{3\left\{s^2 + (\bar{x} - \mu)^2\right\}}{\sigma^2} - 1 + \sum_{i=1}^{k}\frac{C_i}{r}\right] \tag{4.28}$$

and

$$A_i = h(x_i)\left[h(x_i) - z(x_i)\right] \tag{4.29}$$

$$B_i = h(x_i) + z(x_i)A_i \tag{4.30}$$

$$C_i = z(x_i)\left[h(x_i) + B_i\right] \tag{4.31}$$

The estimated parameters are asymptotically normal. The variances of the estimates can be found by inverting the local information matrix,

$$F = \begin{bmatrix} -\frac{\partial^2 L}{\partial \mu^2} & -\frac{\partial^2 L}{\partial \mu \partial \sigma} \\ -\frac{\partial^2 L}{\partial \mu \partial \sigma} & -\frac{\partial^2 L}{\partial \sigma^2} \end{bmatrix} \tag{4.32}$$

After inversion, the variances are

$$F^{-1} = \begin{bmatrix} \operatorname{var}(\hat{\mu}) & \operatorname{cov}(\hat{\mu},\hat{\sigma}) \\ \operatorname{cov}(\hat{\mu},\hat{\sigma}) & \operatorname{var}(\hat{\sigma}) \end{bmatrix} \tag{4.33}$$

Approximate $(1-\alpha)100\%$ confidence intervals for the estimated parameters are

$$\hat{\mu} - K_{\frac{\alpha}{2}}\sqrt{\operatorname{var}(\hat{\mu})} \le \mu \le \hat{\mu} + K_{\frac{\alpha}{2}}\sqrt{\operatorname{var}(\hat{\mu})} \tag{4.34}$$

$$\frac{\hat{\sigma}}{\exp\left(\frac{K_{\frac{\alpha}{2}}\sqrt{\operatorname{var}(\hat{\sigma})}}{\hat{\sigma}}\right)} \le \sigma \le \hat{\sigma}\exp\left(\frac{K_{\frac{\alpha}{2}}\sqrt{\operatorname{var}(\hat{\sigma})}}{\hat{\sigma}}\right) \tag{4.35}$$

where $K_{\frac{\alpha}{2}}$ is the inverse of the standard normal probability density function.

These confidence intervals are approximate but approach exactness as the sample size increases. Confidence intervals for reliability can be found using the expressions

$$\operatorname{var}(\hat{z}) \approx \left(\frac{\operatorname{var}(\hat{\mu}) + \hat{z}^2 \operatorname{var}(\hat{\sigma}) + 2\hat{z}\operatorname{cov}(\hat{\mu},\hat{\sigma})}{\hat{\sigma}^2}\right) \tag{4.36}$$

$$\hat{z} - K_{\frac{\alpha}{2}}\sqrt{\operatorname{var}(\hat{z})} \le z \le \hat{z} + K_{\frac{\alpha}{2}}\sqrt{\operatorname{var}(\hat{z})} \tag{4.37}$$

$$1 - \Phi\left(\hat{z} + K_{\frac{\alpha}{2}}\sqrt{\operatorname{var}(\hat{z})}\right) \le R(x) \le 1 - \Phi\left(\hat{z} - K_{\frac{\alpha}{2}}\sqrt{\operatorname{var}(\hat{z})}\right) \tag{4.38}$$

Confidence intervals for percentiles can be found using the expressions

$$\operatorname{var}(\hat{x}) \approx \operatorname{var}(\hat{\mu}) + \hat{z}^2 \operatorname{var}(\hat{\sigma}) + 2\hat{z}\operatorname{cov}(\hat{\mu},\hat{\sigma}) \tag{4.39}$$

$$\hat{x} - K_{\frac{\alpha}{2}}\sqrt{\operatorname{var}(\hat{x})} \le x \le \hat{x} + K_{\frac{\alpha}{2}}\sqrt{\operatorname{var}(\hat{x})} \tag{4.40}$$

Hazard Plotting

The normal cumulative hazard function is

$$H(x) = -\ln\left[1 - \Phi\left(\frac{x-\mu}{\sigma}\right)\right] \tag{4.41}$$

where $\Phi(x)$ is the standard normal cumulative distribution function.

By rearranging Eq. 4.41, the survival time can be represented as a function of the cumulative hazard function,

$$x = \mu + \sigma\Phi^{-1}\left[1 - e^{-H(x)}\right] \tag{4.42}$$

where $\Phi^{-1}(x)$ is the inverse of the standard normal cumulative distribution function.

It can be seen that by plotting x versus $\Phi^{-1}\left[1 - e^{-H(x)}\right]$, the resulting y-intercept equals μ, and the resulting slope equals σ. The hazard function, $h(x)$, is estimated from the inverse of the reverse rank of the ordered failures, and $H(x)$ is the cumulative of the values of $h(x)$. Censored data points are used to compute ranks but are not included in hazard plots.

An alternative to plotting x versus $\Phi^{-1}\left[1 - e^{-H(x)}\right]$ on conventional graph paper is to plot x versus $H(x)$ on specialized hazard paper. The advantage of hazard paper is that the values of $\Phi^{-1}\left[1 - e^{-H(x)}\right]$ do not have to be computed. Computers have made this technique obsolete.

Example 4.9: Use hazard plotting to determine the parameters of the normal distribution, given the following multiple censored data set. A "c" following an entry indicates censoring.

Time to Fail		
150 c	183 c	235
157 c	209	235
167 c	216 c	248 c
179	217 c	257

Solution: Table 4.3 is constructed to obtain the necessary plotting data.

TABLE 4.3
TABULATIONS FOR A HAZARD PLOT
FOR THE NORMAL DISTRIBUTION

Time to Fail	Reverse Rank	$h(t)$	$H(t)$	$1-e^{-H(t)}$	$\Phi^{-1}\left[1-e^{-H(t)}\right]$
150 c	12				
157 c	11				
167 c	10				
179	9	0.1111	0.1111	0.1052	−1.2527
183 c	8				
209	7	0.1429	0.2540	0.2243	−0.7578
216 c	6				
217 c	5				
235	4	0.2500	0.5040	0.3959	−0.2640
235	3	0.3333	0.8373	0.5671	0.1691
248 c	2				
257	1	1.0000	1.8373	0.8408	0.9976

c = Censored.

Now the five failure times can be plotted against $\Phi^{-1}\left[1-e^{-H(t)}\right]$, as shown in Figure 4.3. The best-fit straight line through the data points was found using linear regression. The y-intercept of the best-fit straight line through the five points provides an estimate of $\mu - 230.3$ in this case, and the slope of the line provides an estimate of $\sigma - 32.9$ in this case.

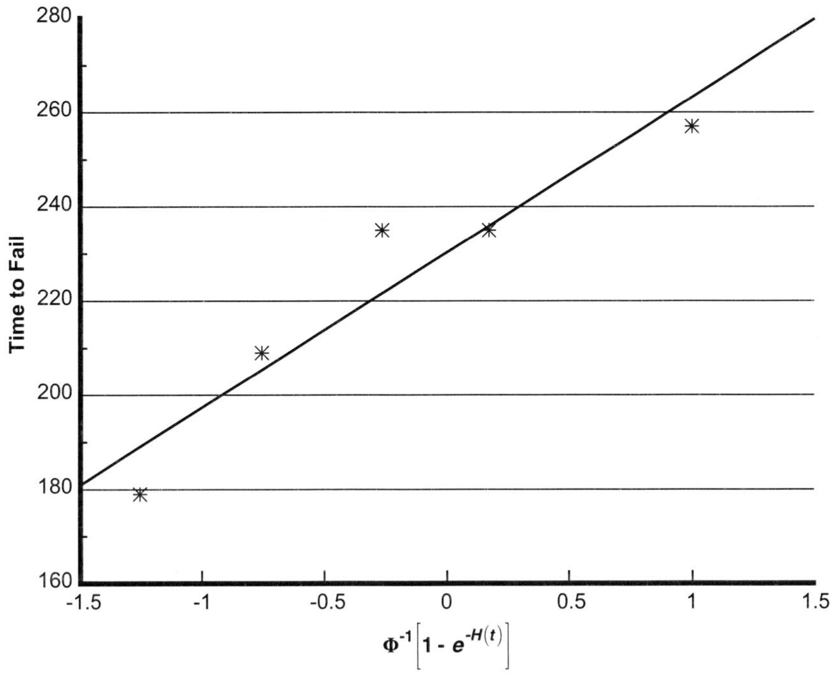

Figure 4.3 *Hazard plot for the normal distribution.*

Probability Plotting

By rearranging the normal cumulative distribution function, a linear expression can be obtained:

$$x = \mu + \sigma\Phi^{-1}\left[F(x)\right] \tag{4.43}$$

where

$F(x)$ is the normal cumulative distribution function

$\Phi^{-1}(x)$ is the inverse of the standard normal cumulative distribution function

It can be seen that by plotting x versus $\Phi^{-1}\left[F(x)\right]$, the resulting y-intercept equals μ, and the resulting slope equals σ. The cumulative distribution function, $F(x)$, is usually estimated from the median rank, but other estimates such as the mean rank and the Kaplan-Meier product limit estimator are also used. Median ranks are estimated using the method detailed in the Exponential Probability Plotting section of this chapter.

An alternative to plotting x versus $\Phi^{-1}\left[F(x)\right]$ on conventional graph paper is to plot x versus $F(x)$ on specialized probability paper. The advantage of probability paper is that the values of $\Phi^{-1}\left[F(x)\right]$ do not have to be computed. Computers have made this technique obsolete.

Example 4.10: Use probability plotting to determine the parameters of the normal distribution, given the following data set. A "c" following an entry indicates censoring.

Time to Fail		
150 c	183 c	235
157 c	209	235
167 c	216 c	248 c
179	217 c	257

Solution: Table 4.4 is constructed to obtain the data for plotting. These data are plotted in Figure 4.4. The slope of this plot is 34.8, which is the estimated value of σ. The y-intercept is 235.3, which is the estimated value of μ.

TABLE 4.4
TABULATIONS FOR A PROBABILITY PLOT FOR NORMAL DATA

Time to Fail	I_i	O_i	Median Rank, $F(x)$	$\Phi^{-1}[F(x)]$
150 c				
157 c				
167 c				
179	1.3000	1.3000	0.0806	–1.4008
183 c				
209	1.4625	2.7625	0.1986	–0.8467
216 c				
217 c				
235	2.0475	4.8100	0.3637	–0.3486
235	2.0475	6.8575	0.5288	0.0723
248 c				
257	3.0713	9.9288	0.7765	0.7605

c = Censored.

Nonparametric confidence limits for reliability can be added to a normal probability plot using 5% and 95% ranks, as explained in the Exponential Probability Plotting section of this chapter. The plotting position for the standard normal inverse of the 5% and 95% ranks is

$$x = \mu + \sigma\Phi^{-1}\left[F(x)\right] \tag{4.44}$$

Note that $F(x)$ is the median rank.

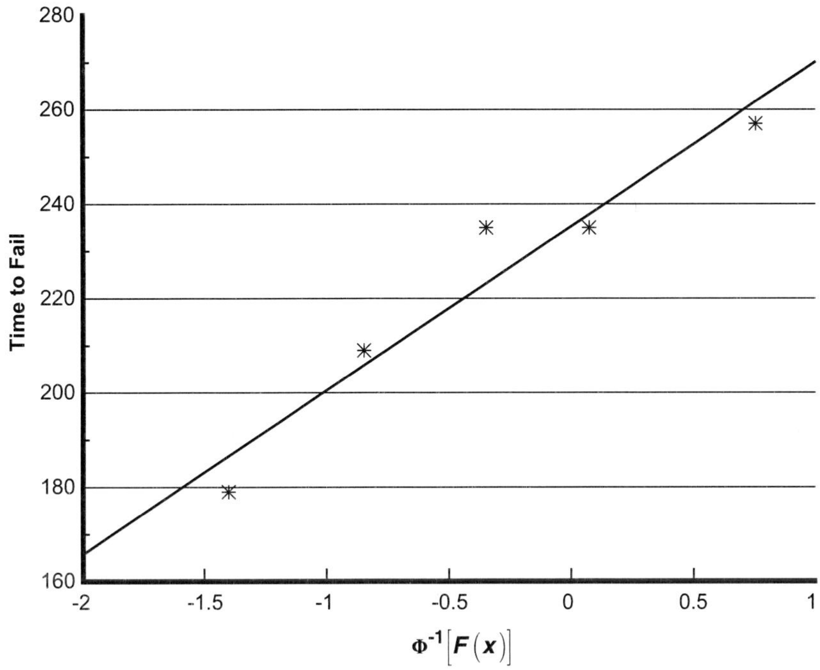

Figure 4.4 *Normal probability plot.*

Lognormal Distribution

The lognormal probability density function is

$$f(x) = \frac{1}{\sigma x\sqrt{2\pi}} \exp\left[-\frac{1}{2}\left(\frac{\ln x - \mu}{\sigma}\right)^2\right], \quad x > 0 \tag{4.45}$$

where

μ is the location parameter

σ is the shape parameter

If x is a lognormally distributed random variable, then $y = \ln(x)$ is a normally distributed random variable. The location parameter is equal to the mean of the logarithm of the data points, and the shape parameter is equal to the standard deviation of the logarithm of the data points. Thus, the lognormal distribution does not have to be dealt with as a separate distribution. By taking the logarithm of the data points, the techniques developed for the normal distribution discussed in the previous section can be used to estimate the parameters of the lognormal distribution.

Weibull Distribution

The Weibull probability density function is

$$f(x) = \frac{\beta(x-\delta)^{\beta-1}}{\theta^{\beta}}\exp\left(\frac{x-\delta}{\theta}\right)^{\beta}, \quad x \geq \delta \tag{4.46}$$

where

β is the shape parameter

θ is the scale parameter

δ is the location parameter

In some cases, a three-parameter Weibull distribution provides a better fit than the two-parameter Weibull distribution. The difference in the two distributions is the location parameter δ, which shifts the distribution along the x-axis. By definition, there is a zero probability of failure for $x < \delta$. Although unusual, the location can be negative; this implies that items were failed prior to testing.

Three methods for estimating the parameters of the Weibull distribution are presented in this section:

1. Maximum likelihood estimation
2. Hazard plotting
3. Probability plotting

Maximum Likelihood Estimation

The following techniques are applicable for the two-parameter and three-parameter Weibull distributions. When using the three-parameter Weibull distribution, replace x with $x - \delta$. The maximum likelihood equations for the Weibull distribution are

$$\frac{1}{r}\sum_{i=1}^{r}\ln\left(x_i\right) = \left[\sum_{i=1}^{n} x_i^{\beta}\ln\left(x_i\right)\right]\left[\sum_{i=1}^{n} x_i^{\beta}\right]^{-1} - \frac{1}{\beta} \tag{4.47}$$

$$\hat{\theta} = \left[\frac{1}{r}\sum_{i=1}^{n} x_i^{\beta}\right]^{\frac{1}{\beta}} \tag{4.48}$$

where

r is the number of failures

n is the total number of data points (both censored and uncensored)

Iterative techniques are required for this expression. The estimated parameters are asymptotically normal. The variances of the estimates can be found by inverting the local information matrix. The local information matrix is

$$F = \begin{bmatrix} -\dfrac{\partial^2 L}{\partial \beta^2} & -\dfrac{\partial^2 L}{\partial \beta \partial \theta} \\ -\dfrac{\partial^2 L}{\partial \beta \partial \theta} & -\dfrac{\partial^2 L}{\partial \theta^2} \end{bmatrix} \tag{4.49}$$

The second partial derivatives of the log-likelihood equation are

$$\frac{\partial^2 L}{\partial \beta^2} = \sum_r \left[-\frac{1}{\beta} - \left(\frac{x_i}{\theta}\right)^\beta \ln^2\left(\frac{x_i}{\theta}\right) \right] + \sum_k \left[-\left(\frac{x_i}{\theta}\right)^\beta \ln^2\left(\frac{x_i}{\theta}\right) \right] \tag{4.50}$$

$$\frac{\partial^2 L}{\partial \theta^2} = \sum_r \left[\frac{\beta}{\theta^2} - \left(\frac{x_i}{\theta}\right)^\beta \left(\frac{\beta}{\theta^2}\right)(\beta+1) \right] + \sum_k \left[-\left(\frac{x_i}{\theta}\right)^\beta \left(\frac{\beta}{\theta^2}\right)(\beta+1) \right] \tag{4.51}$$

$$\frac{\partial^2 L}{\partial \beta \partial \theta} = \sum_r \left\{ -\frac{1}{\theta} + \left(\frac{x_i}{\theta}\right)^\beta \left(\frac{1}{\theta}\right)\left[\beta \ln\left(\frac{x_i}{\theta}\right) + 1\right] \right\} + \sum_k \left\{ \left(\frac{x_i}{\theta}\right)^\beta \left(\frac{1}{\theta}\right)\left[\beta \ln\left(\frac{x_i}{\theta}\right) + 1\right] \right\} \tag{4.52}$$

where

$\sum_r$ represents summation over all failures

$\sum_k$ represents summation over all censored points

The variances of the estimated parameters are

$$F^{-1} = \begin{bmatrix} \operatorname{var}(\beta) & \operatorname{cov}(\beta,\theta) \\ \operatorname{cov}(\beta,\theta) & \operatorname{var}(\theta) \end{bmatrix} \tag{4.53}$$

Approximate $(1-\alpha)100\%$ confidence intervals for the estimated parameters are

$$\frac{\beta}{\exp\left(\dfrac{K_{\frac{\alpha}{2}}\sqrt{\operatorname{var}(\beta)}}{\beta}\right)} \le \beta \le \beta \exp\left(\frac{K_{\frac{\alpha}{2}}\sqrt{\operatorname{var}(\beta)}}{\beta}\right) \tag{4.54}$$

$$\frac{\theta}{\exp\left(\frac{K_{\frac{\alpha}{2}}\sqrt{\operatorname{var}(\theta)}}{\theta}\right)} \leq \theta \leq \theta \exp\left(\frac{K_{\frac{\alpha}{2}}\sqrt{\operatorname{var}(\theta)}}{\theta}\right) \tag{4.55}$$

where $K_{\frac{\alpha}{2}}$ is the inverse of the standard normal probability density function.

These confidence intervals are approximate but approach exactness as the sample size increases. Confidence intervals for reliability can be found using the expressions

$$\exp\left[-\exp\left(u + K_{\frac{\alpha}{2}}\sqrt{\operatorname{var}(u)}\right)\right] \leq R(x) \leq \exp\left[-\exp\left(u - K_{\frac{\alpha}{2}}\sqrt{\operatorname{var}(u)}\right)\right] \tag{4.56}$$

$$u = \beta\left[\ln(x) - \ln(\theta)\right] \tag{4.57}$$

$$\operatorname{var}(u) \approx \beta^2\left[\left(\frac{\operatorname{var}(\theta)}{\theta^2}\right) + \left(\frac{u^2 \operatorname{var}(\beta)}{\beta^4}\right) - \left(\frac{2u \operatorname{cov}(\beta,\theta)}{\beta^2\theta}\right)\right] \tag{4.58}$$

Confidence intervals for percentiles can be found using the expressions

$$e^{y_L} \leq x \leq e^{y_U} \tag{4.59}$$

$$x = \theta\left[-\ln(1-p)\right]^{\frac{1}{\beta}} \tag{4.60}$$

$$y_L = \ln\theta + \frac{\ln\left[-\ln(1-p)\right]}{\beta} - K_\alpha\sqrt{\operatorname{var}(y)} \tag{4.61}$$

$$y_U = \ln\theta + \frac{\ln\left[-\ln(1-p)\right]}{\beta} + K_\alpha\sqrt{\operatorname{var}(y)} \tag{4.62}$$

$$\operatorname{var}(y) = \frac{\operatorname{var}(\theta)}{\theta^2} + \frac{\left\{\ln\left[-\ln(1-p)\right]\right\}^2 \operatorname{var}(\beta)}{\beta^4} - \frac{2\left\{\ln\left[-\ln(1-p)\right]\right\}\operatorname{cov}(\theta,\beta)}{\beta^2\theta} \tag{4.63}$$

Hazard Plotting

The Weibull cumulative hazard function is

$$H(x) = -\ln\left[1 - F(x)\right] \tag{4.64}$$

Replacing $F(x)$ and rearranging gives a linear expression

$$\ln H(x) = \beta \ln x - \beta \ln \theta \tag{4.65}$$

By plotting $\ln H(x)$ versus $\ln x$, the resulting slope (censored points are not plotted) provides an estimate of β. The y-intercept of this plot is an estimate of $\beta \ln \theta$. Thus, θ is estimated from the expression

$$\hat{\theta} = \exp\left(-\frac{y_0}{\hat{\beta}}\right) \tag{4.66}$$

where y_0 is the y-intercept of the hazard plot.

The hazard function $h(x)$ is estimated from the inverse of the reverse rank of the ordered failures; the cumulative hazard function, $H(x)$, is the cumulative of the values of $h(x)$. An alternative to plotting $\ln H(x)$ versus $\ln x$ is to directly plot $H(x)$ versus x on specialized Weibull hazard paper. The advantage of hazard paper is that logarithmic transformations do not have to be computed. Computers have made this technique obsolete.

Example 4.11: Determine the parameters of the Weibull distribution using the multiple censored data in the following table. A "c" following an entry indicates censoring.

Time to Fail	
309 c	229
386	104 c
180	217 c
167 c	168
122	138

Solution: Table 4.5 is constructed to obtain the necessary plotting data.

Now the final two columns of Table 4.5 can be plotted, as shown in Figure 4.5. The slope of the best-fit straight line through the data (found by linear regression) is equal to 2.34 and provides an estimate of β. The y-intercept of the best-fit straight line through the data is –13.004. The estimated scale parameter for the Weibull distribution is

$$\hat{\theta} = \exp\left(-\frac{-13.004}{2.34}\right) = 259.1$$

TABLE 4.5
TABULATIONS FOR A HAZARD PLOT FOR THE WEIBULL DISTRIBUTION

Time to Fail	Reverse Rank	$h(t)$	$H(t)$	$\ln H(t)$	$\ln t$
104 c	10				
122	9	0.1111	0.1111	–2.1972	4.8040
138	8	0.1250	0.2361	–1.4435	4.9273
167 c	7				
168	6	0.1667	0.4028	–0.9094	5.1240
180	5	0.2000	0.6028	–0.5062	5.1930
217 c	4				
229	3	0.3333	0.9361	–0.0660	5.4337
309 c	2				
386	1	1.0000	1.9361	0.6607	5.9558

c = Censored.

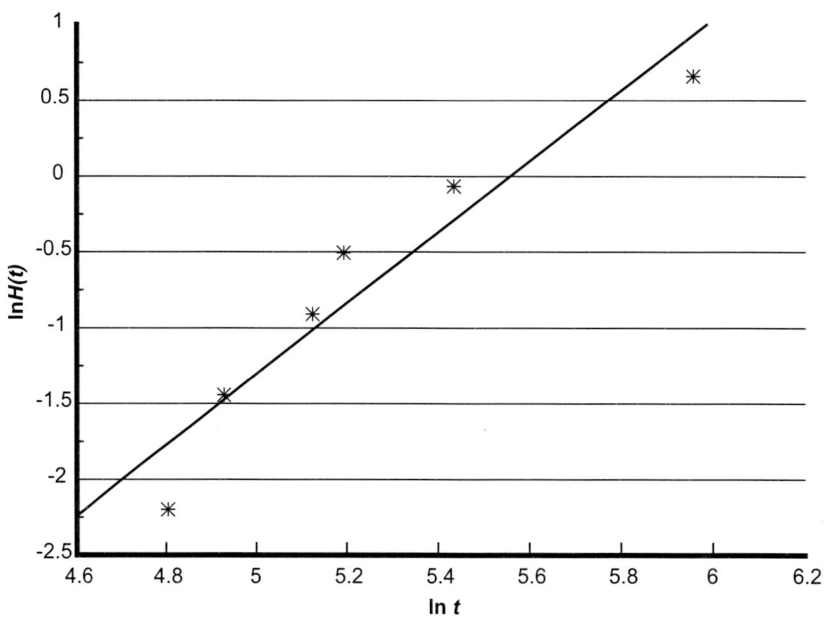

Figure 4.5 *Weibull distribution hazard plot.*

Probability Plotting

By taking the logarithm of the Weibull cumulative distribution function twice and rearranging,

$$\ln\left[\ln\left(\frac{1}{1-F(x)}\right)\right] = \beta \ln x - \beta \ln \theta \tag{4.67}$$

By plotting $\ln\left[\ln\left(\frac{1}{1-F(x)}\right)\right]$ versus $\ln x$ and fitting a straight line to the points, the parameters of the Weibull distribution can be estimated. The slope of the plot provides an estimate of β, and the y-intercept can be used to estimate θ,

$$\hat{\theta} = \exp\left(-\frac{y_0}{\hat{\beta}}\right) \tag{4.68}$$

The cumulative distribution function, $F(x)$, is usually estimated from the median rank, but other estimates such as the mean rank and the Kaplan-Meier product limit estimator are also used. Median ranks are estimated using techniques shown in the Exponential Probability Plotting section of this chapter. Specialized probability paper is available for probability plotting. Using probability paper eliminates the need to transform the data prior to plotting.

Example 4.12: Determine the parameters of the Weibull distribution using probability plotting for the data given in the hazard plotting example, Example 4.11.

Solution: Table 4.6 is constructed to obtain the necessary plotting data.

TABLE 4.6
TABULATIONS FOR A PROBABILITY PLOT FOR THE WEIBULL DISTRIBUTION

Time to Fail	I_i	O_i	Median Rank, $F(x)$	$\ln\left[\ln\left(\frac{1}{1-F(x)}\right)\right]$	$\ln t$
104 c					
122	1.1000	1.1000	0.0769	−2.5252	4.8040
138	1.1000	2.2000	0.1827	−1.6008	4.9273
167 c					
168	1.2571	3.4571	0.3036	−1.0167	5.1240
180	1.2571	4.7143	0.4245	−0.5934	5.1930
217 c					
229	1.5714	6.2857	0.5755	−0.1544	5.4337
309 c					
386	2.3571	8.6429	0.8022	0.4827	5.9558

The last two columns of Table 4.6 are plotted in Figure 4.6. The slope of the best-fit straight line through the data (found using linear regression) is 2.41, which is the estimated value of β. The y-intercept of the best-fit straight line through the data is –13.55. The estimated shape parameter is

$$\hat{\theta} = \exp\left(-\frac{-13.55}{2.41}\right) = 276.6$$

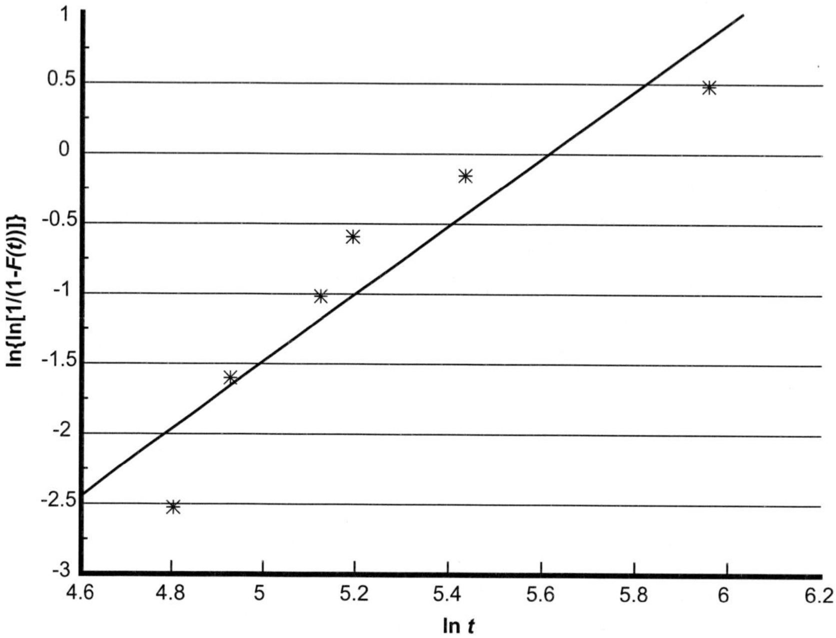

Figure 4.6 *Weibull distribution probability plot.*

Confidence limits can be added to this plot using 5% and 95% ranks as described in the Exponential Probability Plotting section of this chapter. Plotting position for the 5% and 95% ranks are found from the expression

$$x_i = \theta\left[\ln\left(\frac{1}{1-F(x)}\right)\right]^{\frac{1}{\beta}} \tag{4.69}$$

Example 4.13: Repeat Example 4.12, and add 5% and 95% confidence limits.

Solution: The 5% and 95% ranks are computed using the following expression. Because the failure order is not an integer, the 5% and 95% ranks must be interpolated. Using a failure order of 1, from Eq. 4.15 the 5% rank is

$$w_{0.05} = \frac{\frac{1}{10-1+1}}{19.4457 + \frac{1}{10-1+1}} = 0.0051$$

Using a failure order of 2, the 5% rank is

$$w_{0.05} = \frac{\frac{2}{10-2+1}}{5.8211 + \frac{1}{10-2+1}} = 0.0368$$

Interpolating to a failure order of 1.1 gives a 5% rank of 0.0083. All calculations are shown in Table 4.7.

Figure 4.7 shows the probability plot with 5% and 95% confidence limits.

Nonparametric Confidence Intervals

The duration for a bogey test is equal to the reliability requirements being demonstrated. For example, if a test is designed to demonstrate that a component has specific reliability at 100 hours, and the test duration is 100 hours, then the test is designated as a bogey test. Pass-fail tests also are considered bogey tests. A $100(1-\alpha)\%$ nonparametric confidence interval for reliability when bogey testing is

$$\frac{n-r}{n-r+(r+1)F_{\frac{\alpha}{2},2(r+1),2(n-r)}} \le R \le \frac{(n-r+1)F_{\frac{\alpha}{2},2(n-r+1),2r}}{r+(n-r+1)F_{\frac{\alpha}{2},2(n-r+1),2r}} \tag{4.70}$$

where

n is the sample size

r is the number of failures

F_{α,ν_1,ν_2} is the critical value of the F-distribution

TABLE 4.7
PROBABILITY PLOTTING COMPUTATIONS

Time to Fail	Status	Order Increment	Order	Median Rank	5% *F*-Critical	95% *F*-Critical	5% Rank	95% Rank	Interpolated 5% Rank	Interpolated 95% Rank	Rank Plotting Position
104	Censored										
			1		19.4457	0.2863	0.00512	0.25887			
122	Failed	1.1000	1.1000	0.0769					0.0083	0.2724	96.27
			2		5.8211	0.3416	0.03677	0.39416			
138	Failed	1.1000	2.2000	0.1827					0.0469	0.4167	141.17
			3		3.9223	0.3648	0.08726	0.50690			
167	Censored										
168	Failed	1.2571	3.4571	0.3036					0.1160	0.5525	179.80
			4		3.2374	0.3706	0.15003	0.60662			
180	Failed	1.2571	4.7143	0.4245					0.2018	0.6708	214.24
			5		2.9130	0.3632	0.22244	0.69646			
217	Censored										
			6		2.7534	0.3433	0.30354	0.77756			
229	Failed	1.5714	6.2857	0.5755					0.3292	0.7982	256.96
			7		2.6987	0.3089	0.39338	0.84997			
309	Censored		8		2.7413	0.2550	0.49310	0.91274			
386	Failed	2.3571	8.6429	0.8022					0.5656	0.9452	334.54
			9		2.9277	0.1718	0.60584	0.96323			

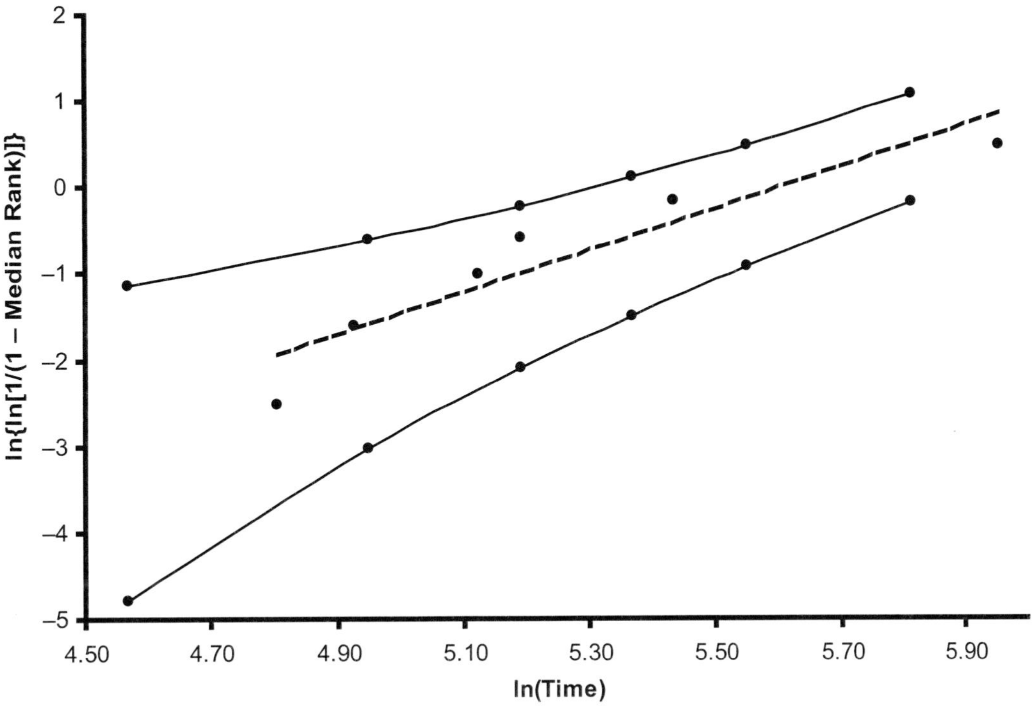

Figure 4.7 *Weibull probability plot with confidence limits.*

Example 4.14: Twenty items are tested, with 3 items failing and 17 items passing. What is the 90% confidence interval (two-sided) for reliability?

Solution: The total sample size, n, is 20; the number of failures, r, is 3. To compute the confidence limits, two critical values from the F-distribution are needed:

$$F_{\frac{\alpha}{2},2(n-r+1),2r} = F_{0.05,36,6} = 3.786$$

$$F_{\frac{\alpha}{2},2(r+1),2(n-r)} = F_{0.05,8,34} = 2.225$$

The confidence interval for reliability is

$$\frac{20-3}{20-3+(3+1)2.225} \leq R \leq \frac{(20-3+1)3.786}{3+(20-3+1)3.786}$$

$$0.656 \leq R \leq 0.958$$

Summary

Several methods are available to estimate distribution parameters. It is recommended to use a graphical method, such as hazard plotting or probability plotting, to obtain a visual goodness-of-fit test. Once the goodness of fit is satisfied, maximum likelihood estimation should be used because it is more accurate than other methods.

Chapter 5

Accelerated Test Plans

Reliability testing is often expensive; expensive prototypes, lengthy tests, and exotic test equipment are not unusual. To deal with time and cost pressures as well as the mathematical complications of censored data, reliability testing statistics have become specialized. Zero-failure testing, sequential testing, accelerated testing, and burn-in are discussed in this chapter.

Mean Time to Fail

Manufacturers often quote mean time to fail as a method of conveying product reliability. Reliability comparisons should never be made based on the mean time to fail. Consider two airplanes. Airplane A has a mean time to fail of 30 hours, and Airplane B has a mean time to fail of 20 hours. You are required to choose one of the airplanes for a 12-hour flight; which do you prefer? The normal reaction is to choose Airplane A. Now consider that the time to fail for both airplanes is normally distributed; the standard deviation for Airplane A is 10 hours, and the standard deviation for Airplane B is 2.5 hours. Do you still prefer Airplane A? The distributions for the time to fail for both airplanes are given in Figure 5.1. The reliability for Airplane A at 12 hours is 96.41%. The reliability for Airplane B at 12 hours is 99.93%. The mean time to fail did not accurately select the airplane with the best reliability at 12 hours.

Test Plan Problems

A reliability test plan requires all of the following:

1. A sample size
2. A test duration
3. Criteria for acceptable performance
4. The number of failures allowed

For example, given a Weibull time-to-fail distribution with a shape parameter of 2, 95% reliability can be demonstrated at 100 hours with a confidence level of 90% by any of the test plans shown in Table 5.1.

The most efficient reliability test plans assume there will be no failures. Of course, this is an unrealistic assumption, but if failure occurs, the item under consideration has not necessarily failed to meet the reliability requirements. The test may be continued until the reliability requirements are met or until it is evident that the reliability requirements will not be met. This defies

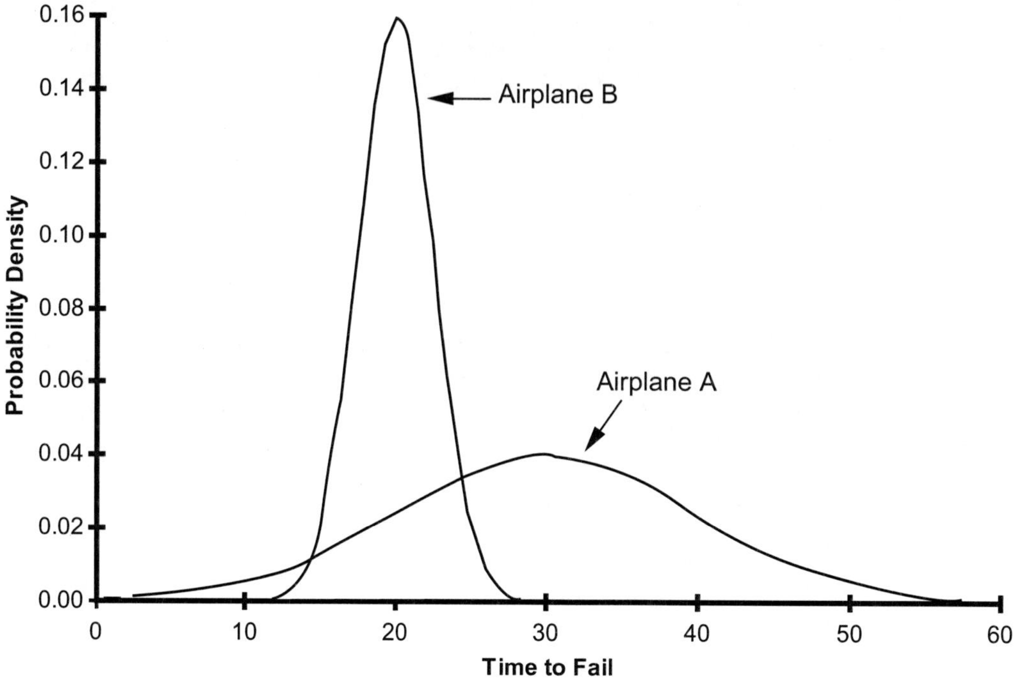

Figure 5.1 *Example of mean time to fail.*

TABLE 5.1
WEIBULL RELIABILITY DEMONSTRATION TEST PLANS

Test Plan	Number of Units Tested	Number of Failures	Test Duration (Hours)
A	45	0	100
B	76	1	100
C	192	5	100
D	318	10	100
E	45	1	131

traditional statistical thinking, which requires test parameters and pass-fail requirements to be determined prior to testing.

Consider a unit with a Weibull time-to-fail distribution having a shape parameter of 2 and a scale parameter of 570 hours. The reliability at 100 hours is

$$R(100) = e^{-\left[\left(\frac{100}{570}\right)^2\right]} = 0.9697$$

Now suppose 45 of these units are tested for 100 hours. To demonstrate 95% reliability at 100 hours of operation with a confidence of 90%, all 45 units must survive for 100 hours. The probability of all 45 units successfully completing the test is

$$P = 0.9697^{45} = 0.2503$$

Although the unit has a reliability greater than 95% at 100 hours, the unit is unable to demonstrate 95% reliability at 100 hours only in slightly more than 25% of the attempts using the designed test plan. Figure 5.2 shows the probability of successfully completing this test plan as a function of the true reliability at 100 hours. As the true reliability increases, the probability of successfully demonstrating 95% reliability increases; as the sample size increases, the probability of successfully demonstrating 95% reliability increases. From Figure 5.2, it can be seen that this item has a 62.89% chance of successfully demonstrating the desired reliability if 318 units are tested.

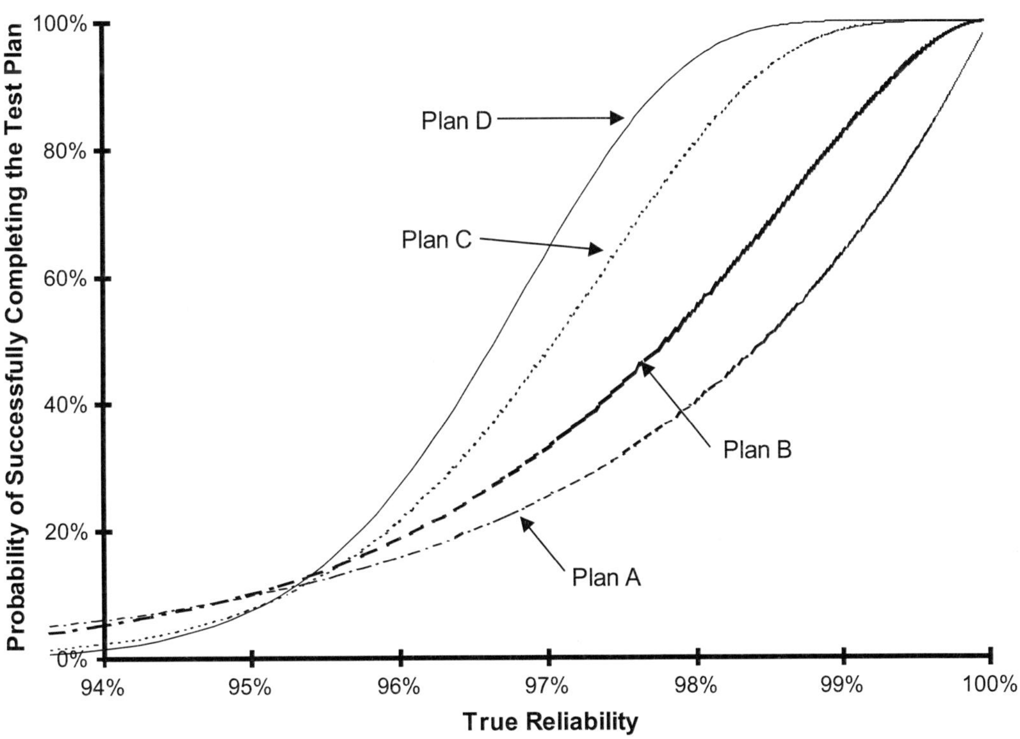

Figure 5.2 *Probability of successful completion of the test plan.*

Why does a unit that exceeds the required reliability fail the reliability demonstration test? Reliability is not demonstrated because the sample size is not large enough. Consider another example: A golfer has a driving distance that is normally distributed with a mean of 250 meters and a standard deviation of 15 meters. This golfer wants to join a league that requires all members to have an average drive of at least 240 meters. This is tested by allowing the golfer to select a number of drives, and after all drives have been hit, if the average drive distance is greater than 240, the golfer is allowed to join the league.

Recall that Type I error is the risk of incorrectly rejecting the null hypothesis, and that Type II error is the risk of incorrectly accepting the null hypothesis. In this example, Type I error is the risk of incorrectly determining that the golfer's average driving distance is less than 240 meters. For this example, the Type II risk is 0; there is no chance of incorrectly determining that the golfer's average driving distance is greater than 240 meters because the average driving distance is greater than 250 meters. The only reason this golfer will not be allowed to join the league is because the golfer's average driving distance is incorrectly evaluated. Figure 5.3 demonstrates the probability of the golfer being allowed to join the league if the driving average is based on a single drive. The probability the golfer is not allowed to join the league is 25.3%. This is the Type I error; the golfer's true average driving distance exceeds 240 meters, but the demonstration test incorrectly determined the true average driving distance was less than 240 meters.

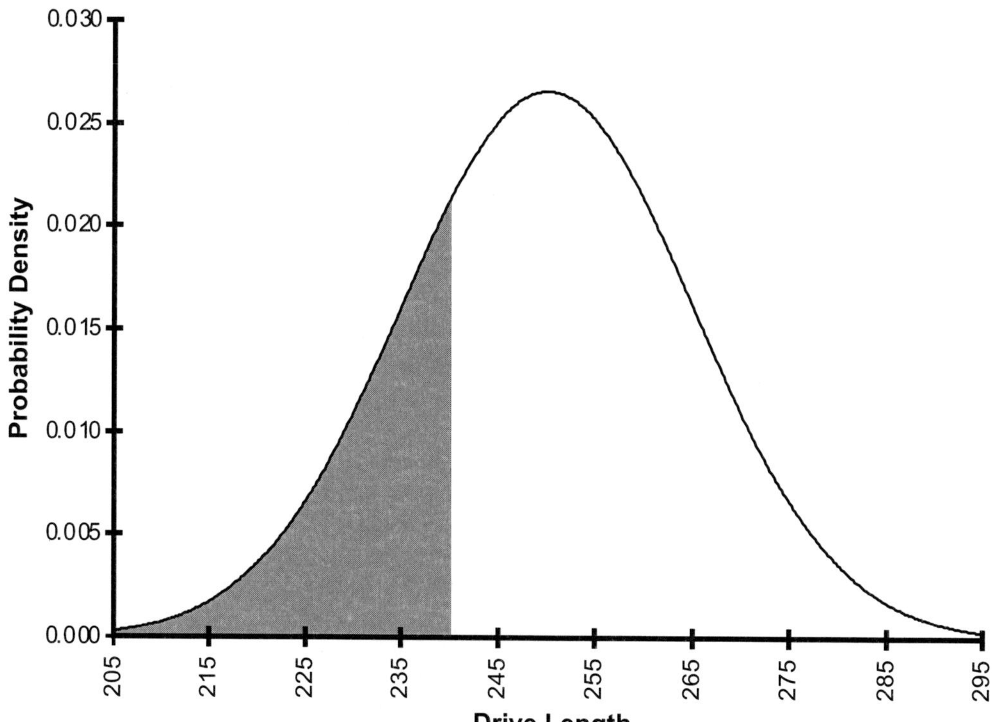

Figure 5.3 *Type I error using a single drive test.*

The distribution of the average n drives has the following properties:

$$\mu = 250$$

$$\sigma = \frac{15}{\sqrt{n}}$$

Figures 5.4 and 5.5 show the effect of sample size on Type I error. By increasing the sample size, the standard deviation of the distribution of the average driving distance is decreased, resulting in a more accurate estimate of the true driving distance. If the golfer is evaluated using the average of 5 drives, there is a 6.8% of failing to meet the requirements to join the league. This error drops to 1.75% when the golfer is evaluated using 10 drives.

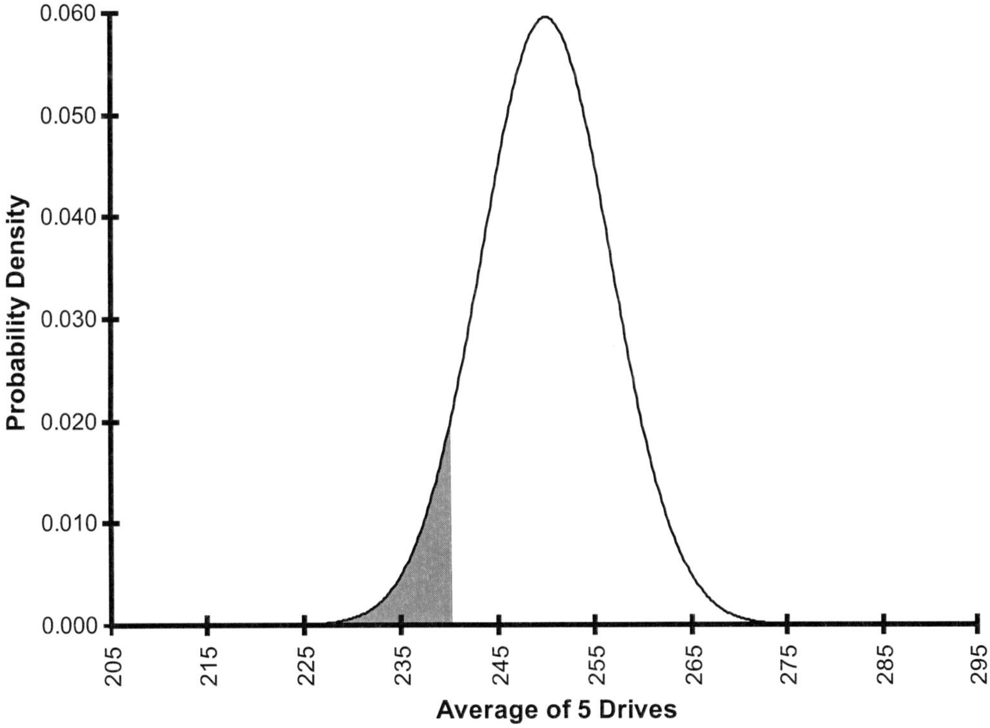

Figure 5.4 *Type I error using a drive test with a sample of 5.*

Figure 5.6 shows the probability of failing to pass the drive test as a function of the number of drives. This figure shows that there is a substantial risk of incorrectly disqualifying the golfer if a small sample size is used, and that as the sample size increases, this error is reduced.

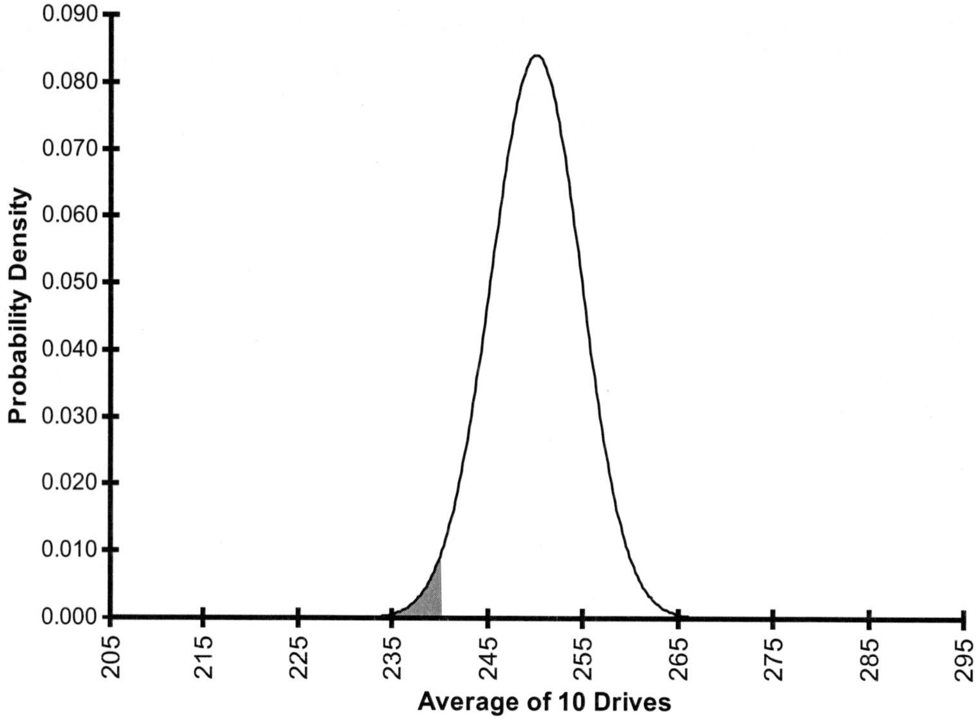

Figure 5.5 *Type I error using a drive test with a sample of 10.*

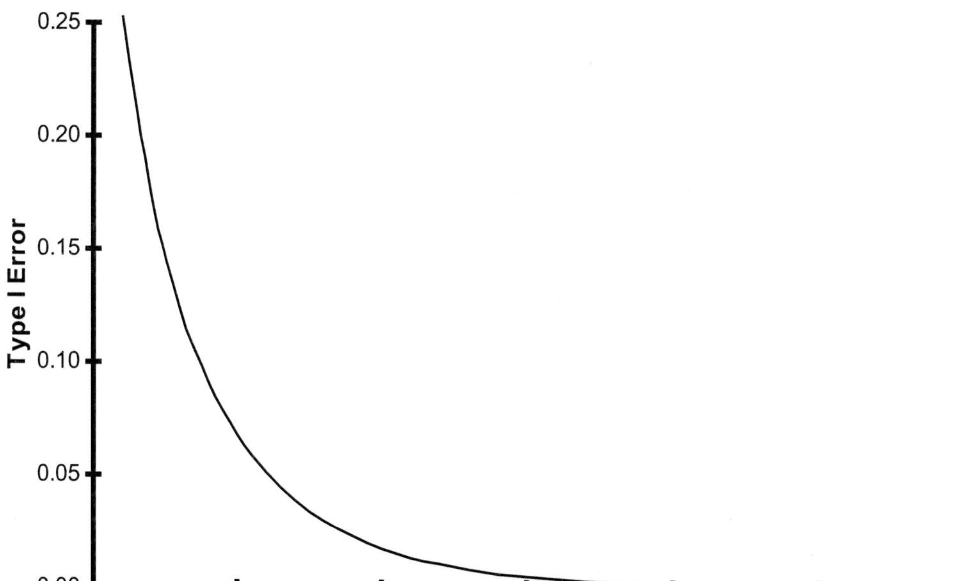

Figure 5.6 *Type I error as a function of sample size.*

Theoretically, this situation is not a problem. If at least 25 drives are required to evaluate each golfer, the error is near 0. However, reliability engineers often do not have this option. Sample sizes of 25 can be too expensive or require too much time. Using a sample size of 2, more than 17% of the evaluations when the golfer exceeds the required mean by 10 meters are incorrectly rejected. If this were a test to evaluate a design or process, the incorrect test result leads to increased cost and delays while the product or process is redesigned. It may be better to place test more samples, knowing that the item under consideration may have been rejected incorrectly, based on a small sample size. This is the concept of sequential testing.

Now consider a golfer who has an average driving distance of 230 meters with a standard deviation of 15 meters. Type I error—the risk of incorrectly determining that the golfer's average driving distance is less than 240 meters—is 0 because the golfer's true driving distance is 230 meters. Type II risk—the probability of incorrectly determining that the golfer's average driving distance is greater than 240—is the same as the Type I risk for the golfer with an average driving range of 250 meters. The only reason this golfer will be allowed to join the league is because the golfer's average driving distance is incorrectly evaluated. Statistical parameters are included in sequential testing to guard against Type I and Type II errors. A spreadsheet titled "TestDesignInteractive.xls" has been included on the accompanying CD to allow further investigation of these situations.

Zero-Failure Testing

There are two types of zero-failure tests: (1) bogey testing, and (2) Bayesian testing.

Bogey Testing

Bogey testing requires the testing duration to be equal to the required life. For example, if 95% reliability is required at 200,000 kilometers of service, then the units being tested will be removed from testing when they fail or when they complete the equivalent of 200,000 kilometers of testing. The sample size required to demonstrate reliability of r with a confidence level of c is

$$n = \frac{\ln(1-c)}{\ln r} \tag{5.1}$$

Example 5.1: A windshield wiper motor must demonstrate 99% reliability with 90% confidence at 2 million cycles of operation. How many motors must be tested to 2 million cycles with no failures to meet these requirements?

Solution: Two hundred and thirty motors must function for 2 million cycles to demonstrate the desired level of reliability, as shown:

$$\frac{\ln(1-0.9)}{\ln 0.99} = 229.1$$

Bayesian Testing

Bogey testing is inefficient. By extending the test duration beyond the required life, the total time on test can often be reduced. When the test duration is extended, it is necessary to make assumptions concerning the shape of the distribution of the time to fail. This is done by assuming a Weibull distribution for time to fail and by assuming a shape parameter. Recall from Chapter 2, Probability Fundamentals, that the Weibull distribution can approximate many other distributions by changing the value of the shape parameter.

Effect of the Shape Parameter

The shape of the distribution has a tremendous effect on the amount of testing required to demonstrate reliability. Consider the histogram of time to fail for a population with a Weibull time-to-fail distribution with a shape parameter of 3.6, as shown in Figure 5.7. Ninety-five percent of the items survived for more than 1 bogey; thus, the reliability at 1 bogey is 95%. This histogram appears to show a normally distributed population because the Weibull distribution is nearly normal when the shape parameter is 3.6. Also, notice that when the shape parameter is 3.6, to demonstrate 95% reliability at 1 bogey, the mean must be at least 2.06 bogeys.

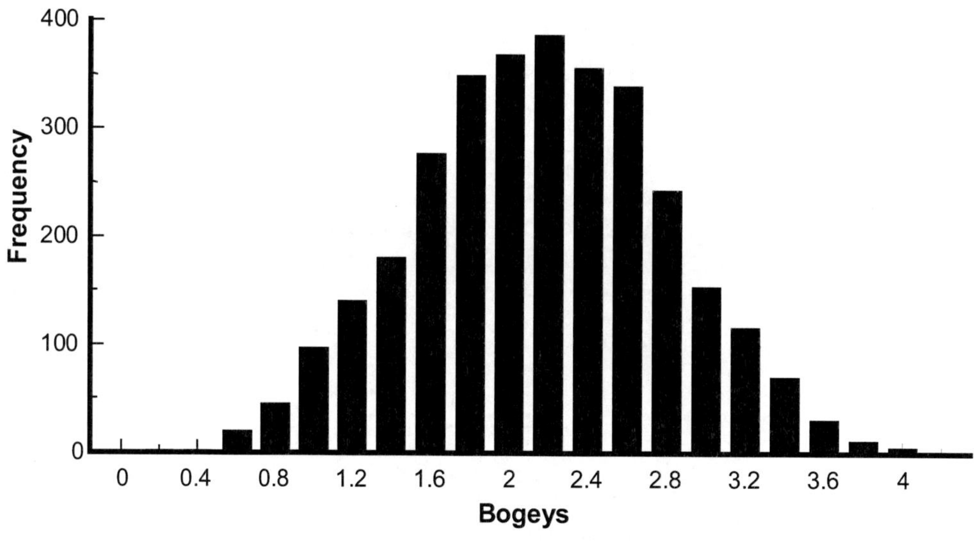

Figure 5.7 *Weibull distributed population with 95% reliability at 1 bogey and a shape parameter of 3.6.*

Figure 5.8 shows a population having a Weibull time-to-fail distribution with a shape parameter of 1. This figure shows a reliability of 95% at 1 bogey and a mean of 19.5 bogeys—nearly 10 times greater than the Weibull distribution with a shape parameter of 3.6. The variance of the distribution increases as the shape parameter decreases. Variance is the equivalent of uncertainty,

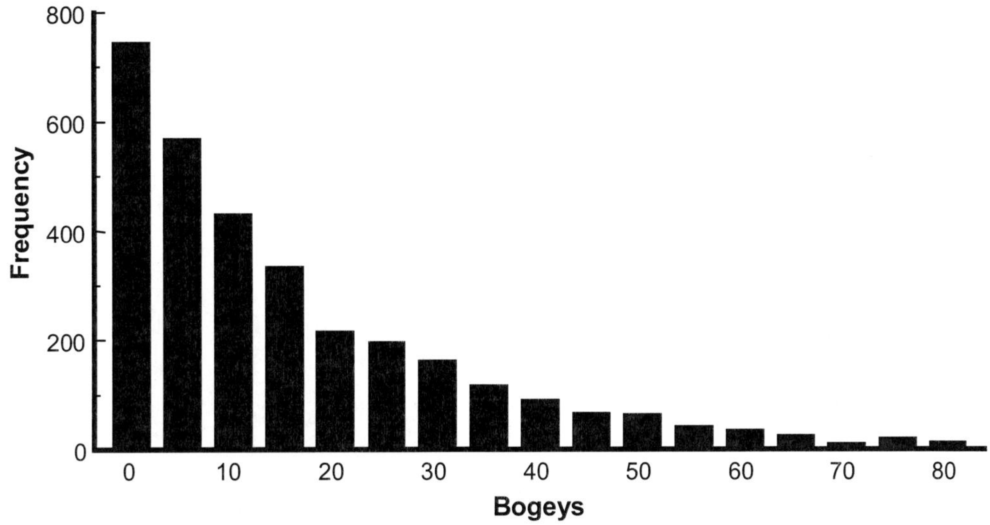

Figure 5.8 Weibull distributed population with 95% reliability at 1 bogey and a shape parameter of 1.0.

and the amount of testing required to demonstrate reliability is dependent on the amount of variance in the population.

Figure 5.9 shows a population having a Weibull time-to-fail distribution with a shape parameter of 1.8. This figure shows a reliability of 95% at 1 bogey and a mean of 5.2 bogeys.

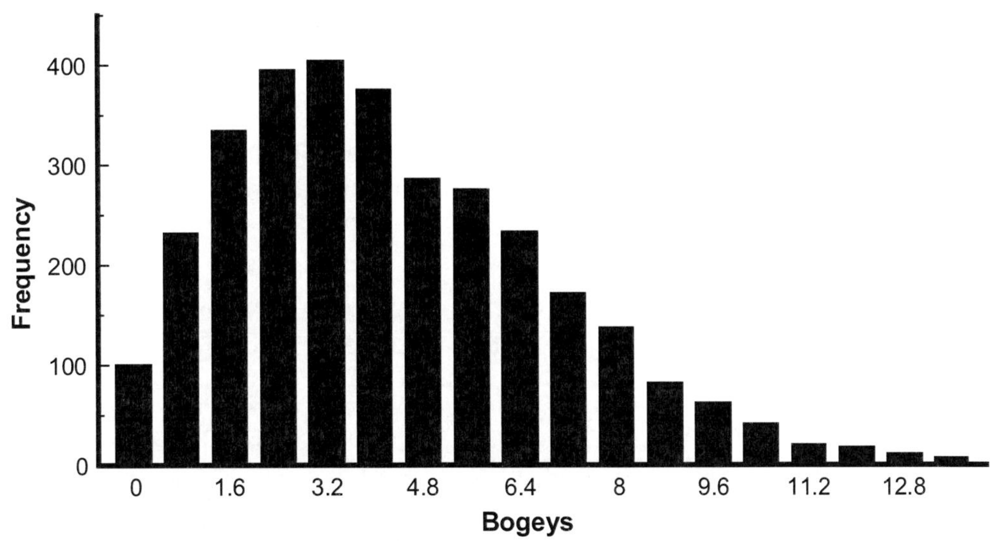

Figure 5.9 Weibull distributed population with 95% reliability at 1 bogey and a shape parameter of 1.8.

Figure 5.10 shows a population having a Weibull time-to-fail distribution with a shape parameter of 8.0. This figure shows a reliability of 95% at 1 bogey and a mean of 1.37 bogeys.

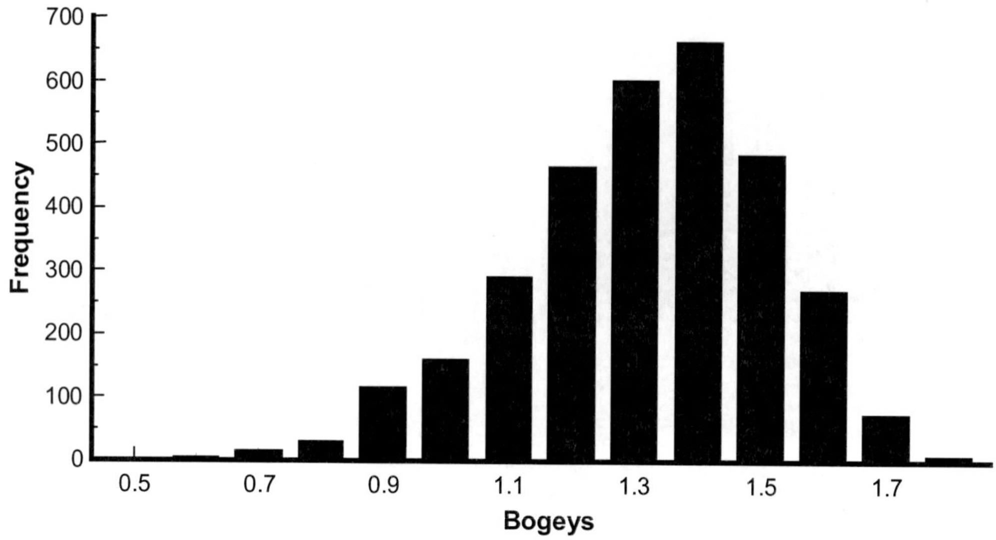

Figure 5.10 *Weibull distributed population with 95% reliability at 1 bogey and a shape parameter of 8.0.*

Estimating the Shape Parameter

From Figures 5.7 through 5.10 inclusive, it can be seen that the assumed shape greatly affects the testing requirements. How is the shape parameter estimated? The shape parameter is governed by the physics of failure. Some phenomena are skewed right, while others are skewed left or have no skewness. In general, the more the failure mode is a result of mechanical wear, the larger the shape parameter will be. The shape parameter is usually stable and tends to remain constant. For example, the shape parameter for master cylinders will be similar for master cylinders for small cars, large trucks, and for different designs because the physics of how the master cylinder fails is similar.

The best way to estimate the shape parameter is through prior testing. Many automotive companies require some testing to failure to allow the shape parameter to be determined. Keep detailed records of all tests, and build a database of shape parameters. It is recommended to use the lower 90% confidence limit for the shape parameter because of the magnitude the shape parameter has on test requirements. In lack of any prior knowledge, data sources are available on the Internet, or the shape parameter can be estimated based on the knowledge of the physics of failure.

Be careful when estimating the shape parameter for electronics. Many sources state the shape parameter for electronics is 1.0 because there is no mechanical wear in electronics. Electronic modules located in environmentally harsh conditions, such as under the hood of an automobile or in an airplane, fail as a result of mechanical wear. The extreme vibration, temperature cycling, and, in some cases, contaminants cause mechanical failures. It is not uncommon to have shape parameters greater than 8.0 for electronic components.

Determining Test Parameters

The statistical properties for Bayesian testing with the Weibull distribution are based on the exponential distribution. If the parameter t follows a Weibull distribution, then the parameter t^{β} is exponentially distributed. The lower $(1-\alpha)$ confidence limit for reliability is

$$R_{L,\alpha}(t) = e^{\left[-\left(\frac{t}{\theta_{L,\alpha}}\right)^{\beta}\right]} \tag{5.2}$$

where

$$\theta_{L,\alpha} = \left(\frac{2\sum_{i=1}^{n} t_i^{\beta}}{\chi^2_{\alpha,d}}\right)^{\frac{1}{\beta}} \tag{5.3}$$

n is the number of units tested, both failed and surviving, and $\chi^2_{\alpha,d}$ is the critical value of the chi-square distribution with significance of α (0.05 for a confidence level of 95%) and d degrees of freedom. For failure truncated testing, d is equal to $2r$, where r is the number of failed units. For time truncated testing, d is equal to $2r + 2$.

Example 5.2: Fourteen units are placed on a test stand. The first unit fails after 324 hours of testing, the second unit fails after 487 hours of testing, the third unit fails after 528 hours of testing, and the remaining 11 units are removed from testing. Given a Weibull time-to-fail distribution with a shape parameter of 2.2, what is the lower 90% confidence limit for reliability at 400 hours?

Solution: Because testing was suspended at the time of the last failure, this is failure truncated testing. With failure truncated testing and 3 failures, the degrees of freedom for the chi-square distribution are $2(3) = 6$. The critical value of the chi-square distribution given 6 degrees of freedom and a significance of 10% is 10.64. This value can be found from Appendix A or using the Microsoft® Excel expression

$$= \text{CHIINV}(1\text{-}0.9, 6)$$

The lower 90% confidence limit for the mean of the transformed data is

$$\theta_{L,\alpha} = \left(\frac{2(12{,}872{,}524.77)}{10.64}\right)^{\frac{1}{2.2}} = 797.3$$

The lower 90% confidence limit for the reliability at 400 hours is

$$R_{L,0.10}(400) = e^{\left[-\left(\frac{400}{797.3}\right)^{2.2}\right]} = 0.80311$$

Example 5.3: Fourteen units are placed on a test stand. The first unit fails after 324 hours of testing, the second unit fails after 487 hours of testing, the third unit fails after 528 hours of testing, and the remaining 11 units are removed from testing after 550 hours without failing. Given a Weibull time-to-fail distribution with a shape parameter of 2.2, what is the lower 90% confidence limit for reliability at 400 hours?

Solution: Because testing was not suspended at the time of the last failure, this is time truncated testing. With time truncated testing and 3 failures, the degrees of freedom for the chi-square distribution are $2(3)+2=8$. The critical value of the chi-square distribution given 8 degrees of freedom and a significance of 10% is 13.36. This value can be found from Appendix A or using the Microsoft Excel expression

$$= \text{CHIINV}(1\text{-}0.9, 8)$$

The lower 90% confidence limit for the mean of the transformed data is

$$\theta_{L,\alpha} = \left(\frac{2(13,882,130.48)}{13.36}\right)^{\frac{1}{2.2}} = 744.1$$

The lower 90% confidence limit for the reliability at 400 hours is

$$R_{L,0.10}(400) = e^{\left[-\left(\frac{400}{744.1}\right)^{2.2}\right]} = 0.7747$$

Reliability tests are often designed in a two-step procedure: (1) How many test stands are available? and (2) What is the test duration, given the number of test fixtures? For a sample of n units, the required test duration to demonstrate a reliability of R at time t with a confidence level of $1-\alpha$ assuming no failures is

$$T = \left(\frac{t}{[-\ln R]^{\frac{1}{\beta}}}\right)\left(\frac{-\ln\alpha}{n}\right)^{\frac{1}{\beta}} \tag{5.4}$$

Example 5.4: How long must 20 units be tested to demonstrate 99% reliability with 80% confidence at 200 hours of operation, given a Weibull time-to-fail distribution with a shape parameter of 3.5 and assuming no failures?

Solution: The required test duration is

$$T = \left(\frac{200}{\left[-\ln(0.99) \right]^{\frac{1}{3.5}}} \right) \left(\frac{-\ln(0.2)}{20} \right)^{\frac{1}{3.5}} = 362 \text{ hours}$$

For a test duration of T, the required sample size to demonstrate a reliability of R at time t with a confidence level of $1 - \alpha$ assuming no failures is

$$n = \frac{-\ln \alpha}{\left[T \left(\frac{(-\ln R)^{\frac{1}{\beta}}}{t} \right) \right]^{\beta}} \tag{5.5}$$

Example 5.5: How many units must be tested for 300 hours to demonstrate 99% reliability with 80% confidence at 200 hours of operation, given a Weibull time-to-fail distribution with a shape parameter of 3.5 and assuming no failures?

Solution: The required sample size is

$$n = \frac{-\ln(1 - 0.80)}{\left[300 \left(\frac{(-\ln 0.99)^{\frac{1}{3.5}}}{200} \right) \right]^{3.5}} = 38.7 \Rightarrow 39$$

Guidelines for Zero-Failure Test Planning

When designing reliability demonstration tests, a balance must be maintained between test duration and sample size. Figure 5.11 shows the effect of sample size on test duration for several shape parameters of the Weibull distribution.

Figure 5.12 shows the effect of sample size on the total test time (sample size × test duration) for several values of the Weibull shape parameter.

From Figure 5.12, it can be seen that when the shape parameter is less than 1.0, the total test time decreases as the sample size increases. If the shape parameter is equal to 1.0, the total test time is unaffected by the sample size. If the shape parameter is greater than 1.0, the total

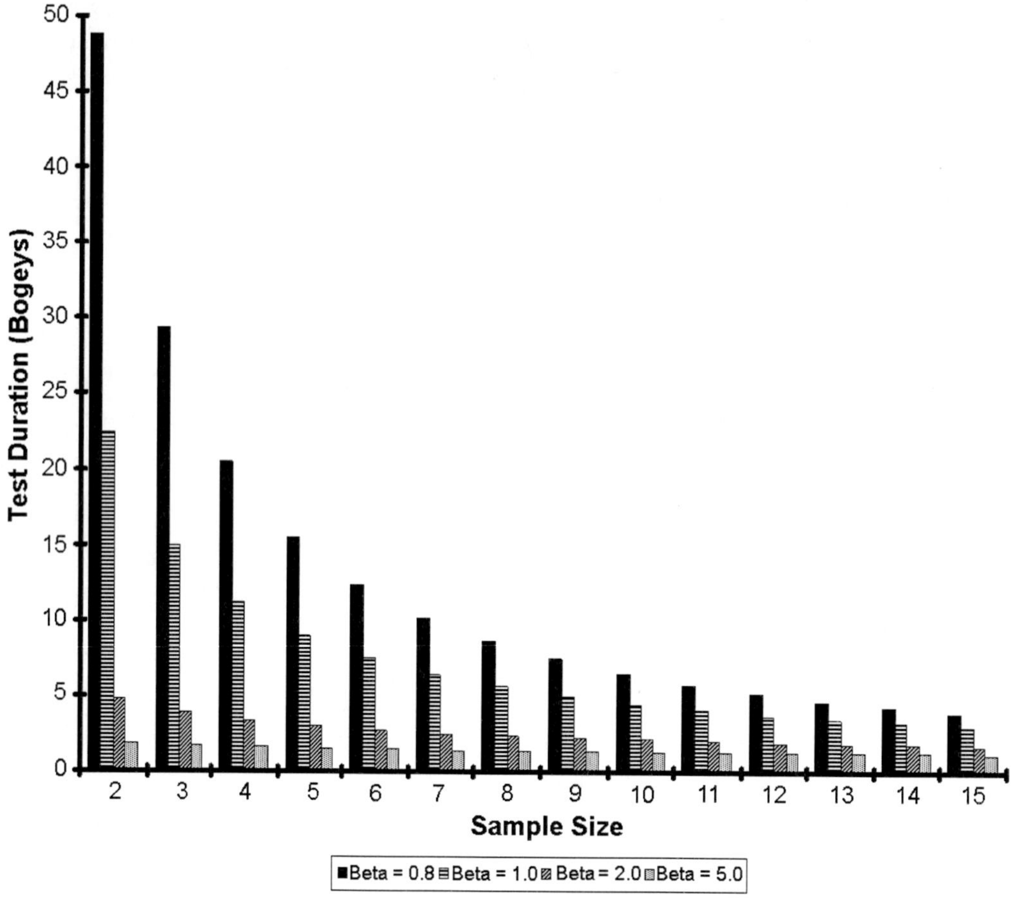

Figure 5.11 *The effect of sample size on test duration.*

test time decreases as the sample size decreases, with the total test time at a minimum when only one sample is tested. Testing one sample is not recommended; if possible, a minimum of four samples should be used to gain an estimate of the variance in the population.

Example 5.6: A major Tier 1 automotive supplier produces DC motors for windshield wiper systems. Depending on the size of the vehicle, the windshield configuration, and other factors, the size and configuration of the motor changes, resulting in a unique motor for each windshield program. Because each program has a unique motor, each motor design must be validated for reliability. The shape parameter results for the five most recent motor development programs are given in Table 5.2. For the next windshield wiper motor program, how many units must be tested and for what test duration to demonstrate 95% reliability with 90% confidence at 1.5 million cycles?

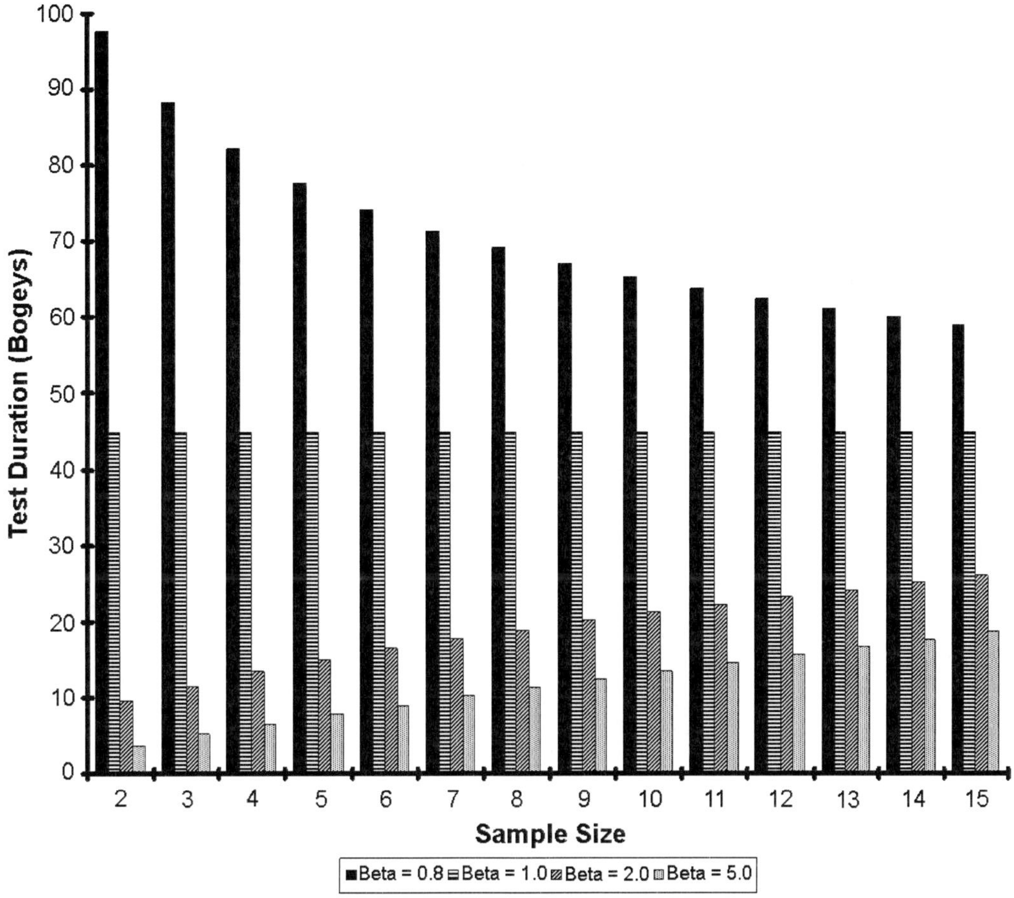

Figure 5.12 *The effect of sample size on total test time.*

TABLE 5.2
MOTOR TEST SHAPE PARAMETER RESULTS

Program	β	Lower 95% Confidence Limit for β
Chrysler—Compact Car	7.41	4.78
Chrysler—Large Car	8.63	5.64
Chrysler—Sport Utility Vehicle	7.04	4.22
Chrysler—Large Car	8.93	6.20
General Motors—Midsize Car	7.99	5.31

Solution: The most efficient test design will assume no failures. This is not a realistic assumption, but it provides the starting point for an efficient test. What value should be used for the shape parameter? A smaller shape parameter requires more testing; thus, the conservative approach is to err on the low side when determining the value of the shape parameter. When data from only one test are available, a safe approach is to use the lower 95% confidence limit for the estimated shape parameter. Another approach is to use the smallest estimated shape parameter—in this case, 7.04. Estimating the shape parameter to use from previous data is not an exact science. Caution should be used, and use of a conservative approach is recommended. In this case, the Tier 1 supplier and the automotive company agreed to use a shape parameter of 7.04 to design the test. Eight test stands were available for testing, which results in a test duration of

$$T = \left(\frac{1.5}{\left[-\ln(0.95)^{\frac{1}{7.04}} \right]} \right) \left(\frac{-\ln(0.1)}{8} \right)^{\frac{1}{7.04}} = 1.92 \text{ million cycles}$$

One of the 8 units failed after 1.82 million cycles of testing. This did not result in the design failing to demonstrate the required reliability. There were two options at this point: (1) redesign the motor and retest, or (2) continue testing the remaining 7 motors until the required reliability is demonstrated. The second choice can be risky; the motor may not be able to demonstrate the required reliability with additional testing.

To demonstrate, the additional 7 motors must survive for a duration of

$$T = \left[\frac{X^2 \left[\frac{1.5}{\left[-\ln(0.95) \right]^{\frac{1}{7.04}}} \right]^{7.04} - 2(1.82)}{2(7)} \right]^{\frac{1}{7.04}} = 2.088 \text{ million cycles}$$

where X^2 is the critical value of the chi-square distribution with a significance of 10% $(1 - 90\%$ confidence$)$ and 4 degrees of freedom $(2r + 2)$, which is 7.7794. The Tier 1 supplier and the automotive company agreed to continue testing the remaining 7 motors rather than abandon the motor design. The remaining 7 motors all survived for 2.088 million cycles.

Sequential Testing

Example 5.6 is an example of sequential testing. Reliability tests often are too lengthy or expensive to begin testing with a sample size large enough to make a decision with a high degree of confidence. Conventional testing requires determining a sample size, null hypothesis, and alternative hypothesis prior to testing. An alternative is to begin with a small sample size (often only a single sample is used) and to make decisions concerning the test hypothesis as test data are collected.

With sequential testing, decisions are based on the likelihood ratio, $\frac{L_{1,n}}{L_{0,n}}$, where

$$L_{k,n} = \prod_{i=1}^{n} f\left(x_i \backslash \theta_k\right) \tag{5.6}$$

where

$f\left(x\right)$ is the probability density function of the variable x

θ is the parameter being tested

The hypotheses for a sequential test are

$$H_0\text{:}\ \ \theta = \theta_0$$

$$H_1\text{:}\ \ \theta = \theta_1$$

Sequential test plans have a Type I error (i.e., the probability of rejecting the null hypothesis when it is true) of α when $\theta = \theta_0$, and a Type II error (i.e., the probability of accepting the null hypothesis when it is not true) that is less than or equal to β when $\theta_1 \leq \theta$.

Pass-Fail Testing

In many instances, the outcome of a reliability test is pass or fail for each of the individual items being tested. This situation can be modeled by the binomial distribution, and for an acceptance test, the hypotheses are

$$H_0\text{:}\ \ p \leq p_0$$

$$H_1\text{:}\ \ p > p_0$$

where

p is the probability of failure for an item

p_0 is the probability of failure; the probability of accepting H_0 is $1 - \alpha$

where α is the Type I error or the producer's risk.

As each item is tested, the decision to continue testing, accept the null hypothesis, or reject the null hypothesis is determined from the expression

$$z = \left(\frac{p_1}{p_0}\right)^y \left(\frac{1 - p_1}{1 - p_0}\right)^{n-y} \tag{5.7}$$

where

p_1 is the probability of failure; the probability of accepting H_0 is β

β is the Type II error or the consumer's risk

n is the total number of trials

y is the total number of failures

The test is continued if $A < z < B$, where

$$A = \frac{\beta}{1 - \alpha} \tag{5.8}$$

$$B = \frac{1 - \beta}{\alpha} \tag{5.9}$$

The null hypothesis is accepted if $z < A$ and rejected if $z > B$. If neither of these conditions is met, testing continues.

Example 5.7: A sequential test was conducted to test the hypotheses

$$H_0\colon\ p \le 0.02$$

$$H_1\colon\ p > 0.02$$

The level of producer's risk for the test was 0.05, the level of consumer's risk was 0.1, and $p_1 = 0.1$. The first 3 items tested were successful, and the 4th and 5th items tested failed. Determine the status of the test after the results of each trial were known.

Solution: The values of A and B are

$$A = \frac{0.1}{1 - 0.05} = 0.10526$$

$$B = \frac{1 - 0.1}{0.05} = 18.0$$

The values of z after each trial, and the resulting decisions, are as follows:

Trial	z	Decision
1	0.9184	Continue
2	0.8434	Continue
3	0.7746	Continue
4	3.873	Continue
5	19.364	Reject H_0

It is often useful to represent acceptance tests graphically. The test described in Example 5.7 can be defined as

Accept H_0 if $c < y$

Reject H_0 if $d > y$

Continue testing if $c < y < d$

where

$$c = \frac{n \ln\left(\frac{1-p_0}{1-p_1}\right)}{\ln\left[\left(\frac{p_1}{p_0}\right)\left(\frac{1-p_0}{1-p_1}\right)\right]} - \frac{\ln\left(\frac{1-\alpha}{\beta}\right)}{\ln\left[\left(\frac{p_1}{p_0}\right)\left(\frac{1-p_0}{1-p_1}\right)\right]} \tag{5.10}$$

$$d = \frac{n \ln\left(\frac{1-p_0}{1-p_1}\right)}{\ln\left[\left(\frac{p_1}{p_0}\right)\left(\frac{1-p_0}{1-p_1}\right)\right]} + \frac{\ln\left(\frac{1-\beta}{\alpha}\right)}{\ln\left[\left(\frac{p_1}{p_0}\right)\left(\frac{1-p_0}{1-p_1}\right)\right]} \tag{5.11}$$

Example 5.8: For the test described in Example 5.7, suppose 81 items were tested with the 4th, 18th, 37th, and 73rd items failing, while all other items were successful. Show this test graphically.

Solution: Figure 5.13 shows this sequential test.

From Figure 5.13, it can be seen that the fewest number of trials required to reject the null hypothesis is 2 (i.e., 2 consecutive failures with no successes). It also can be seen from Figure 5.13

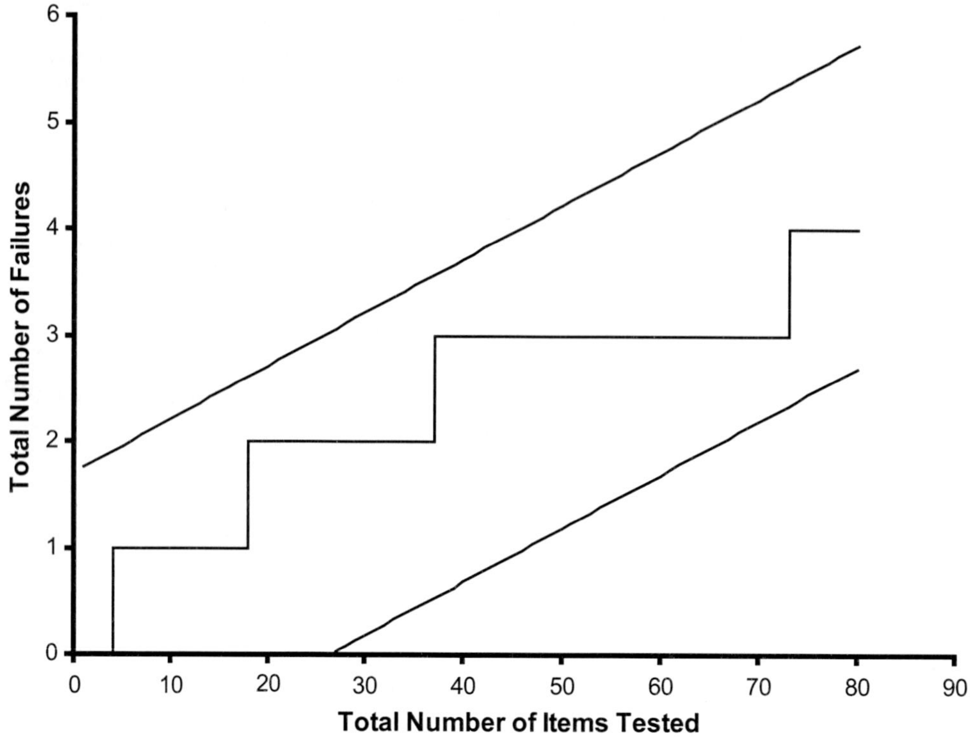

Figure 5.13 *Sequential testing with the binomial distribution.*

that the fewest number of trials (all trials being successful) required to accept the null hypothesis is 27. Mathematically, the fewest number of trials required to reject the null hypothesis is

$$f_R = \frac{\ln \beta}{\ln\left[\left(\frac{p_1}{p_0}\right)\left(\frac{1-p_0}{1-p_1}\right) - \ln\left(\frac{1-p_0}{1-p_1}\right)\right]} \tag{5.12}$$

The fewest number of trials required to accept the null hypothesis is

$$f_A = \frac{\ln A}{\ln\left(\frac{1-p_1}{1-p_0}\right)} \tag{5.13}$$

The discriminatory power of a sampling plan is shown by its operating characteristic curve (OC curve). An OC curve displays the probability of accepting the null hypothesis versus the fraction defective. For binomial sequential sampling, the probability of accepting the null hypothesis is

$$P_a(p) = \frac{B^h - 1}{B^h - A^h} \tag{5.14}$$

where

p is the true fraction defective

h is an arbitrary term used to compute p (h usually ranges from –3 to 3) using the expression

$$p = \frac{1 - \left(\frac{1 - p_1}{1 - p_0}\right)^h}{\left(\frac{p_1}{p_0}\right)^h - \left(\frac{1 - p_1}{1 - p_0}\right)^h} \tag{5.15}$$

Example 5.9: Construct the OC curve for the sampling plan described in Example 5.8.

Solution: First, a value of –3 is arbitrarily assigned to h, and p is computed. The resulting value of p is then used to compute P_a. The value of h is then increased, and the procedure is repeated until enough values are obtained to construct a graph. Table 5.3 contains the information required to construct an OC curve. Figure 5.14 shows this OC curve.

TABLE 5.3
OC CURVE COMPUTATIONS

h	p	P_a
–3.0	0.227	0.001
–2.6	0.201	0.003
–2.2	0.175	0.007
–1.8	0.149	0.017
–1.4	0.124	0.042
–1.0	0.100	0.100
–0.6	0.078	0.224
–0.2	0.059	0.436
0.2	0.043	0.683
0.6	0.030	0.863
1.0	0.020	0.950
1.4	0.013	0.983
1.8	0.008	0.995
2.2	0.005	0.998
2.6	0.003	0.999
3.0	0.002	1.000

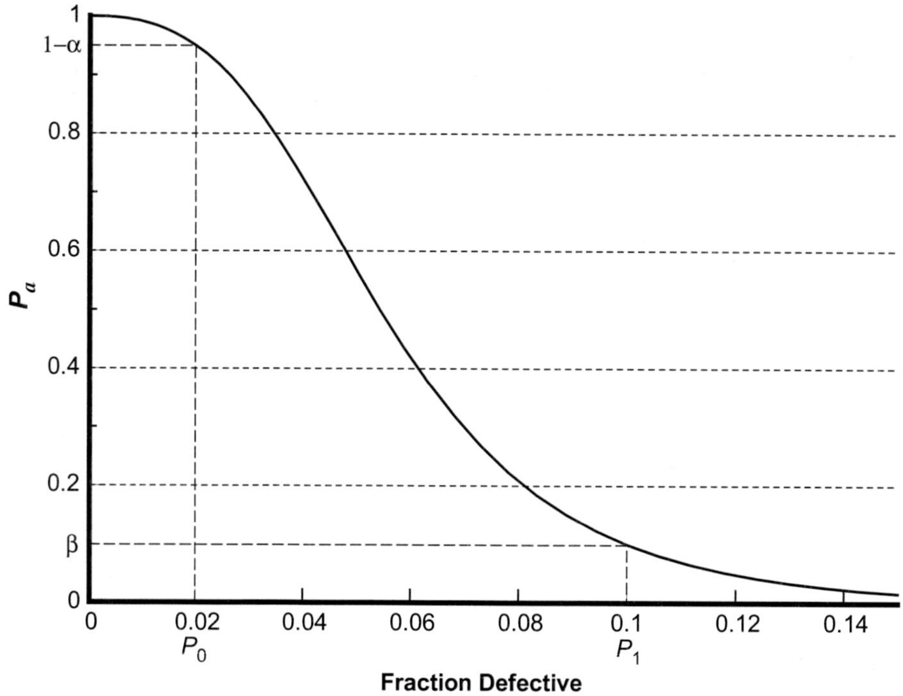

Figure 5.14 *Sequential sampling OC curve.*

Another characteristic that is valuable when planning a sequential test is the expected number of trials required to reach a decision. This value is a function of the true fraction defective and is

$$E(p) = \frac{P_a(p)\ln A + \left[1 - P_a(p)\right]\ln B}{p\ln\left(\frac{p_1}{p_0}\right) + (1-p)\ln\left(\frac{(1-p_1)}{1-p_0}\right)} \tag{5.16}$$

This function reaches a maximum between p_0 and p_1 and is relatively flat in this region.

Example 5.10: Determine the expected number of trials required to reach a decision for the sampling plan described in Example 5.8.

Solution: Figure 5.15 shows the expected number of trials required to reach a decision for the sampling plan described in Example 5.8.

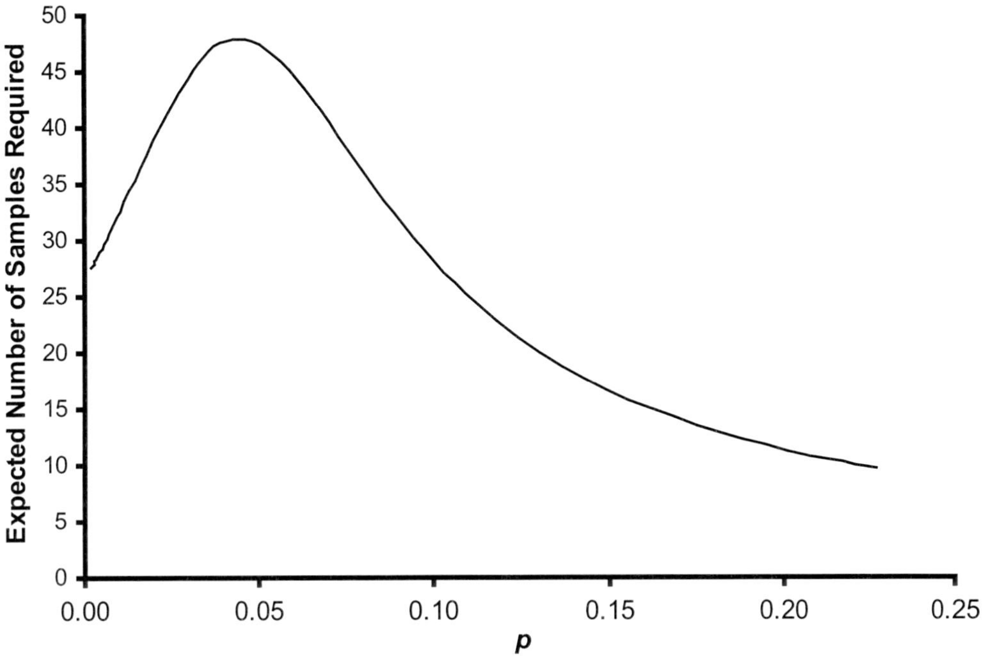

Figure 5.15 Expected number of samples required to make a decision.

Exponential Distribution

Testing with continuous data, such as time to fail or pressure to burst, provides more information than a simple pass or fail designation and reduces testing time. Assuming an exponential time-to-fail distribution, a sequential test can be conducted on the cumulative time to fail,

$$T = \sum_{i=1}^{n} t_i$$

where

t_i is the test time accumulated for the ith unit

n is the total number of units tested

The test hypotheses are

$$H_0\colon \lambda < \lambda_0$$

$$H_1\colon \lambda > \lambda_0$$

where λ is the failure rate of the exponential distribution.

The null hypothesis is rejected if

$$T < \frac{r \ln\left(\dfrac{\lambda_1}{\lambda_0}\right) + \ln\left(\dfrac{\alpha}{1-\beta}\right)}{\lambda_1 - \lambda_0} \tag{5.17}$$

where

r is the number of failures

λ_0 is the failure rate such that the probability of accepting H_0 is $1 - \alpha$

λ_1 is the failure rate such that the probability of accepting H_0 is β

The null hypothesis is accepted if

$$T > \frac{n \ln\left(\dfrac{\lambda_1}{\lambda_0}\right) - \ln\left(\dfrac{\beta}{1-\alpha}\right)}{\lambda_1 - \lambda_0} \tag{5.18}$$

Testing continues if

$$\frac{r \ln\left(\dfrac{\lambda_1}{\lambda_0}\right) + \ln\left(\dfrac{\beta}{1-\alpha}\right)}{\lambda_1 - \lambda_0} < T < \frac{r \ln\left(\dfrac{\lambda_1}{\lambda_0}\right) - \ln\left(\dfrac{\alpha}{1-\beta}\right)}{\lambda_1 - \lambda_0} \tag{5.19}$$

Example 5.11: An accelerated test is needed to demonstrate a failure rate of 0.001 with a producer's risk of 10%. The test should have a 20% chance of passing if the true failure rate is greater than 0.005.

Solution: The parameters for this test are:

$\lambda_0 = 0.001$

$\lambda_1 = 0.005$

$\alpha = 10\%$

$\beta = 20\%$

Figure 5.16 shows this sequential test.

Weibull Distribution

The Bayesian testing described for zero-failure test plans can be expanded to sequential test plans. Again, this assumes a Weibull distribution for the variable of interest with a known or assumed shape parameter. The test hypotheses for the Bayesian Weibull sequential test are

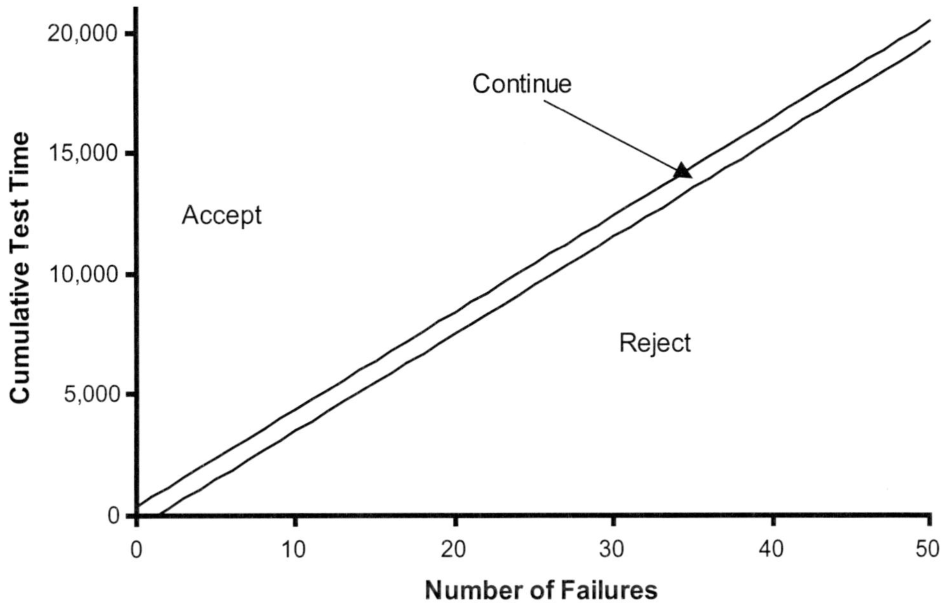

Figure 5.16 *Exponential sequential test.*

$$H_0\text{: } \theta > \theta_0$$

$$H_1\text{: } \theta < \theta_0$$

where

θ is the scale parameter of the Weibull distribution

θ_0 is the scale parameter such that the probability of accepting H_0 is $1 - \alpha$

The desired reliability has been demonstrated if

$$\sum_{i=1}^{n} t_i^{\beta} > nb + d \tag{5.20}$$

and the item has failed to meet reliability requirements if

$$\sum_{i=1}^{n} t_i^{\beta} < nb - c \tag{5.21}$$

where

$$a = \frac{1}{\theta_1^{\beta}} - \frac{1}{\theta_0^{\beta}} \tag{5.22}$$

$$b = \frac{(\beta - 1)\ln\left(\dfrac{\theta_0}{\theta_1}\right)}{a} \tag{5.23}$$

$$c = \frac{\ln\left(\dfrac{1-\tau}{\alpha}\right)}{a} \tag{5.24}$$

$$d = \frac{\ln\left(\dfrac{1-\alpha}{\tau}\right)}{a} \tag{5.25}$$

and θ_1 is the value of the scale parameter that yields a τ probability of accepting H_0. The term τ is used to represent the consumer's risk because the standard term, β, is used to represent the shape parameter of the Weibull distribution.

Example 5.12: A component is required to be 95% reliable after 12.96 weeks. Create a sequential test with a 5% chance of rejecting a component that meets the reliability requirement, while having a 10% chance of accepting a component with reliability less than 89.4% after 12.96 weeks. The time to fail for the component being evaluated is known to follow a Weibull distribution with a shape parameter of 2.2.

Solution: The requirements for reliability must be transformed into required scale parameters for the Weibull distribution. This is done by algebraic manipulation of the Weibull reliability function

$$R(x) = e^{-\left(\frac{t}{\theta}\right)^{\beta}} \Rightarrow \theta = \frac{t}{\left\{-\ln\left[R(t)\right]\right\}^{\frac{1}{\beta}}} \tag{5.26}$$

The scale parameter for a Weibull distribution with a shape parameter of 2.2 having 95% reliability at 12.96 million cycles is

$$\theta_0 = \frac{12.96}{\left[-\ln(0.95)\right]^{\frac{1}{2.2}}} = 50 \text{ weeks}$$

The scale parameter for a Weibull distribution with a shape parameter of 2.2 having 95% reliability at 12.96 million cycles is

$$\theta_1 = \frac{12.96}{\left[-\ln(0.894)\right]^{\frac{1}{2.2}}} = 35 \text{ weeks}$$

The remaining calculations required to construct the test plan are

$$a = \frac{1}{35^{2.2}} - \frac{1}{50^{2.2}} = 0.00021799$$

$$b = \frac{(2.2-1)\ln\left(\frac{50}{35}\right)}{0.00021799} = 1{,}963.4$$

$$c = \frac{\ln\left(\frac{1-0.01}{0.05}\right)}{0.00021799} = 13{,}259$$

$$d = \frac{\ln\left(\frac{1-0.05}{0.01}\right)}{0.00021799} = 10{,}327$$

Figure 5.17 describes this test plan.

Example 5.13: What is the status of the test described in Example 5.12 if the testing shown in Table 5.4 has been completed?

TABLE 5.4
WEIBULL SEQUENTIAL TEST EXAMPLE

Component	Test Time (Weeks)	Status
A	14.8	Suspended
B	15.1	Suspended
C	7.7	Suspended
D	28.0	Suspended

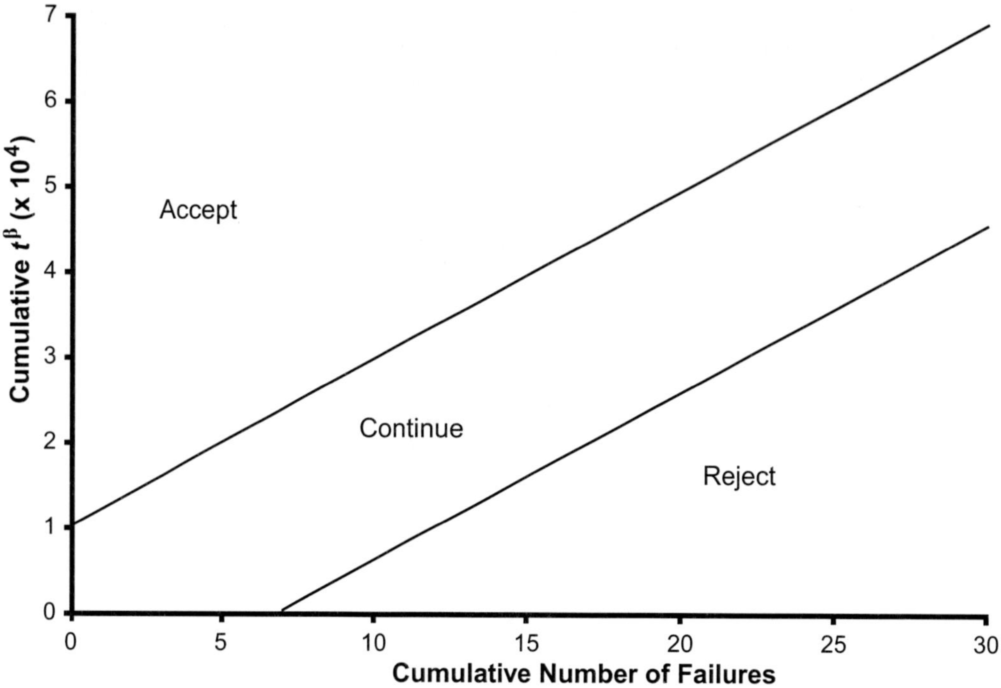

Figure 5.17 *Weibull sequential testing example.*

Solution: The decision parameter is weeks to fail raised to the shape parameter of the Weibull distribution $\left(t^{\beta}\right)$. With no failures, the null hypothesis cannot be rejected, and from Figure 5.17, the component must accumulate a t^{β} value greater than 10,327 to demonstrate the required reliability. Table 5.5 shows the computations for the cumulative value of t^{β}.

TABLE 5.5
WEIBULL SEQUENTIAL TESTING EXAMPLE COMPUTATIONS

Component	Test Time (Weeks)	Status	t^{β}	$\sum t^{\beta}$
A	14.8	Suspended	375	375
B	15.1	Suspended	392	768
C	7.7	Suspended	89	857
D	28.0	Testing	1,527	2,384

With the cumulative value of t^β equal to 2,384, the test is in the continue region.

Example 5.14: For the test described in Examples 5.12 and 5.13, how long must Component D survive to demonstrate the desired reliability?

Solution: The required cumulative value of t^β is 10,327. Components A, B, and C accumulated a total of 857 for t^β. This means that t^β for Component D must be

$$t^\beta = 10{,}327 - 857 = 9{,}470$$

Thus, the required test time for Component D is

$$t = \left(t^\beta\right)^{\frac{1}{\beta}} = 9{,}470^{\frac{1}{2.2}} = 64.2 \text{ weeks}$$

Randomization of Load Cycles

Environmental and stress conditions occur randomly, or randomly within seasons, when used by consumers. This is sometimes overlooked in laboratory testing where testing may begin with low stress and increase the level of stress as testing progresses, completing the test with the highest stress. Because the level of stress randomly occurs when the product is in the field, the test data cannot accurately predict field reliability. To accurately model field reliability, the stress levels applied in the laboratory must accurately match the application rate in the field. In most cases, this requires a random application of stress level.

Distribution-specific random numbers can be generated by setting the cumulative distribution function equal to a unit, uniform random number and taking the inverse of the function. A unit, uniform random number follows a uniform distribution between 0 and 1. The function =*rand*() in Microsoft Excel generates unit, uniform random numbers, and the function *RND* in *Basic* generates unit, uniform random numbers.

Example 5.15: Derive a random number generator for the Weibull distribution.

Solution: For the Weibull distribution,

$$r = 1 - e^{-\left(\frac{x}{\theta}\right)^{\beta}}$$

$$1 - r = e^{-\left(\frac{x}{\theta}\right)^{\beta}}$$

$$\ln(1 - r) = -\left(\frac{x}{\theta}\right)^{\beta}$$

$$\left[-\ln(1 - r)\right]^{\frac{1}{\beta}} = \frac{x}{\theta}$$

$$x = \theta\left[-\ln(1 - r)\right]^{\frac{1}{\beta}}$$

$$x = \theta\left[-\ln(r)\right]^{\frac{1}{\beta}}$$

The last step of the solution is possible because the term $1 - r$ is a unit, uniform distribution that is identical to the distribution characteristics of r.

Table 5.6 gives random number generators for several distributions.

TABLE 5.6
DISTRIBUTION-SPECIFIC RANDOM NUMBER GENERATORS

Distribution	Random Number Generator
Exponential	$x = -\frac{1}{\lambda}\ln r$
Normal	$x = \left[\sqrt{-2\ln r_1}\cos(2\Pi r_2)\right]\sigma + \mu^{*}$
Lognormal	$x = \exp\left[\sqrt{-2\ln r_1}\cos(2\Pi r_2)\right]\sigma + \mu^{*}$
Weibull	$x = \theta(-\ln r)^{\frac{1}{\beta}}$

* Two unit, uniform random numbers are required to generate one distribution-specific random variable.

Example 5.16: Figure 5.18 shows a histogram of the pressure recorded in the fluid transportation system on several vehicles over a one-year time period; a total of 10,624 pressure readings were recorded.* A function is needed for the test controller to randomly generate pressure cycles that follow this histogram.

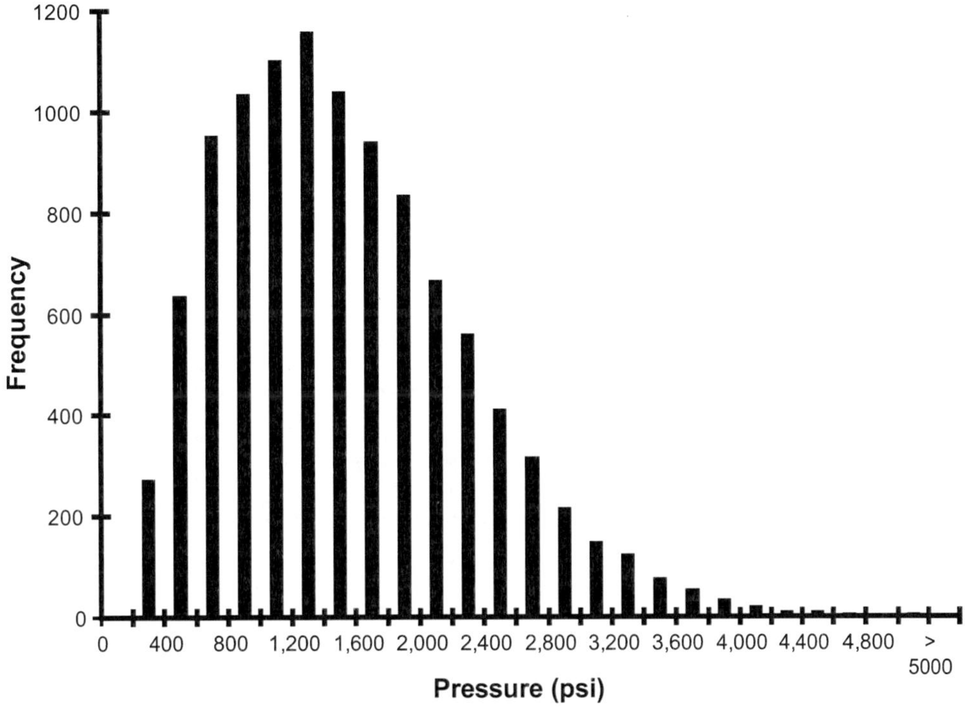

Figure 5.18 *Fluid transportation system pressure histogram.*

Solution: Figure 5.19 shows a Weibull probability plot for this data. The Weibull parameters computed using this plot are

$$\beta = 1.8$$

$$\theta = 1{,}505$$

The function used to randomly generate pressure cycles is

$$p = 1{,}505(-\ln r)^{\frac{1}{1.8}}$$

---

* The data used to create this histogram are given in the file Chapter5.xls on the accompanying CD.

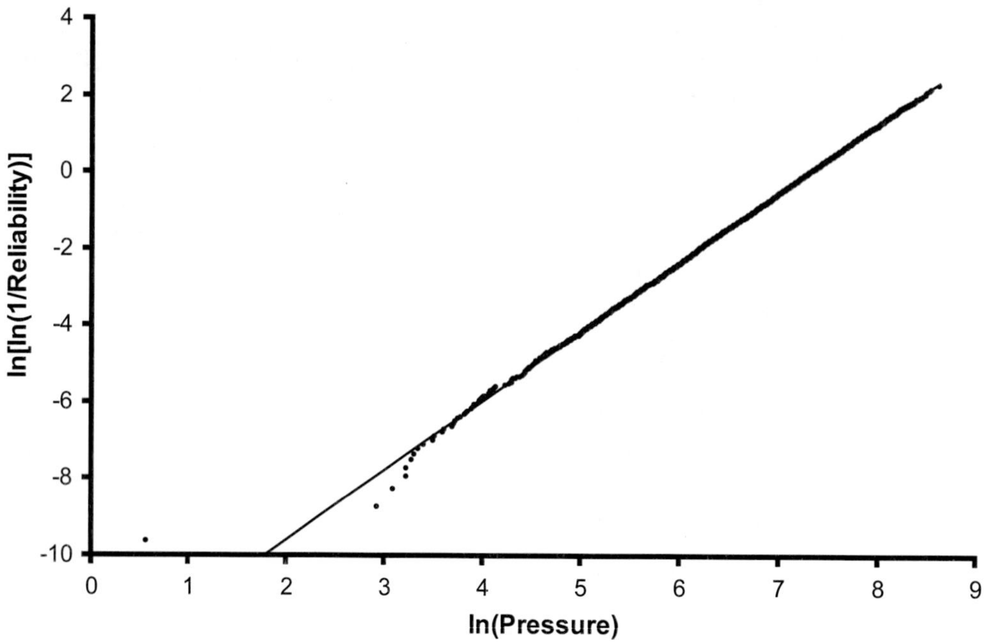

Figure 5.19 *Weibull probability plot for pressure.*

For some environmental factors, such as temperature, it is impossible or prohibitively expensive to completely randomize the stress levels. For these cases, it is recommended to cycle the stress level as often as feasible during the test.

Reliability Growth*

Reliability growth is the improvement in reliability over a period of time due to changes in product design or the manufacturing process. It occurs by surfacing failure modes and implementing effective corrective actions. Reliability growth management is the systematic planning for reliability achievement as a function of time and other resources, and controlling the ongoing rate of achievement by reallocation of these resources based on comparisons between planned and assessed reliability values.

The following benefits can be realized by the utilization of reliability growth management:

- **Finding Unforeseen Deficiencies**—The initial prototypes for a complex system with major technological advances will invariably have significant reliability and performance deficiencies that could not be foreseen in the early design stage. This is also true of prototypes

* Portions of this section were taken from the *1999 AMSAA Reliability Growth Handbook*. (This handbook is available on the accompanying CD.)

that are simply the integration of existing systems. Unforeseen problems are the norm in achieving seamless interoperation and interfacing among systems that have already been developed. Reliability growth testing will surface these deficiencies.

- **Designing in Improvement Through Surfaced Problems**—Even if some potential problems can be foreseen, their significance might not. Prototypes are subjected to a development–testing program to surface those problems that drive the failure rate, so that the necessary improvements in system design can be made. The ensuing system reliability and performance characteristics will depend on the number and effectiveness of these fixes. The ultimate goal of the development test program is to meet the system reliability and performance requirements.
- **Reducing the Risk of Final Demonstration**—Experience has shown that programs that rely simply on a final demonstration alone to determine compliance with the reliability requirements do not, in many cases, achieve the reliability objectives within the allocated resources. Emphasis on reliability performance prior to the final demonstration using quantitative reliability growth could substantially increase the chance of passing or even replace a final demonstration.
- **Increasing the Probability of Meeting Objectives**—This can be achieved by setting interim reliability goals to be met during the development-testing program and the necessary allocation and reallocation of resources to attain these goals. A comprehensive approach to reliability growth management throughout the development program organizes this process.

Reliability Growth Process

Reliability growth is the result of an iterative design process. As the design matures, it is investigated to identify actual or potential sources of failures. Further design effort is then spent on these problem areas. The design effort can be applied to either product design or manufacturing process design. The iterative process can be visualized as the feedback loop shown in Figure 5.20.

Figure 5.20 illustrates the three essential elements involved in achieving reliability growth:

- Detection of failure modes
- Feedback of problems identified
- A redesign effort based on problems identified

Figure 5.20 illustrates the growth process and associated management processes in a skeleton form. This type of illustration is used so that the universal features of these processes may be addressed. The representation of an actual program or program phase may be considerably more detailed. This detailing may include specific inputs to and outputs from the growth process, additional activity blocks, and more explicit decision logic blocks.

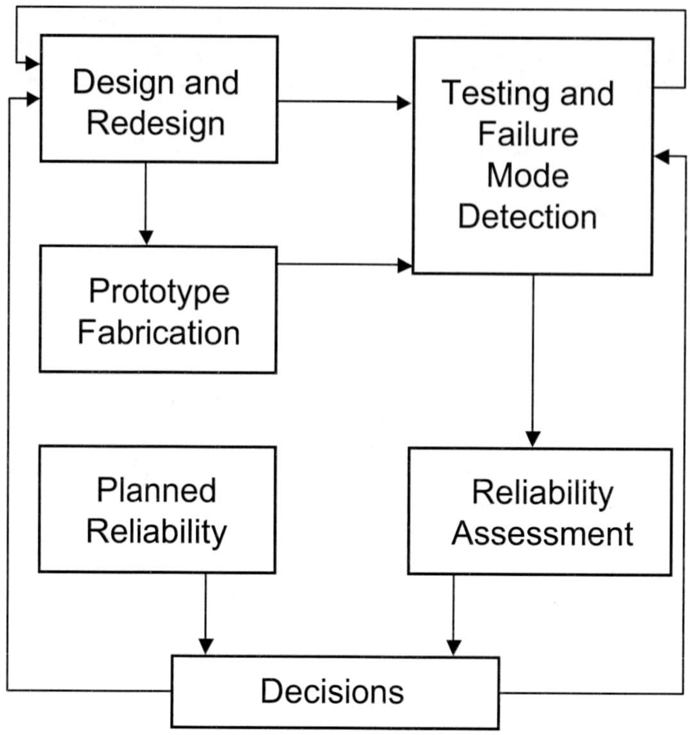

Figure 5.20 *Reliability growth feedback model.*

Reliability Growth Models

Reliability growth models can be used to estimate the amount of time required to obtain a reliability goal or to predict reliability. These growth models apply to software, hardware, and processes. There are several reliability growth models; the two discussed in this section are the Duane model and the Army Material Systems Analysis Activity (AMSAA) model.

Duane Model

J.T. Duane was one of the first to introduce a reliability growth model. He empirically derived the expression shown in Eq. 5.27, which assumes a non-homogeneous Poisson process,

$$\theta_c = \alpha t^{\beta} \tag{5.27}$$

where

θ_c is the cumulative mean time between failures

β is the slope parameter

α is the initial mean time to fail

By taking the logarithm of Eq. 4.57, the following linear expression is found:

$$\ln(\theta_C) = \ln\alpha + \beta\ln t \tag{5.28}$$

The parameters of the Duane model can be found by plotting Eq. 5.28. The current or instantaneous mean time to fail is

$$\theta_i(t) = \frac{\alpha t^{\beta}}{(1-\beta)} \tag{5.29}$$

Example 5.17: Ten electronic throttle bodies were each tested for 250,000 cycles. After each failure, the root cause was determined, and design changes were made on all 10 throttle bodies. Table 5.7 gives the failure data. Determine the Duane model parameters. Predict the testing duration to obtain a mean cycles to fail of 75,000 cycles.

TABLE 5.7
RELIABILITY GROWTH EXAMPLE DATA

Cumulative Cycles (Thousands)	Cumulative Failures	Cumulative Mean Cycles to Fail (Thousands)
250	12	20.83
500	20	25.00
750	27	27.78
1000	31	32.26
1250	33	37.88

Solution: The parameters of the Duane growth model are determined by plotting the logarithm of the cumulative failures versus the logarithm of cumulative test cycles. Table 5.8 contains the information required for the graph.

Figure 5.21 shows a graph of the last two columns of Table 5.8.

The slope of the straight line in Figure 5.21 is $\beta = 0.353$. The y-intercept of the graph is 1.05. The initial mean time to fail is

$$\alpha = e^{y\text{-intercept}} = 2.864 \text{ (thousands of cycles)}$$

The Duane reliability growth model for the instantaneous mean cycles to fail is

$$\theta_i(t) = \frac{2.864t^{0.353}}{(1-0.353)}$$

TABLE 5.8
COMPUTATIONS FOR DUANE GROWTH MODEL EXAMPLE

Cumulative Cycles (Thousands)	Cumulative Failures	Cumulative Mean Cycles to Fail (Thousands)	Logarithm of Cumulative Cycles (Thousands)	Logarithm of Cumulative Mean Cycles to Fail (Thousands)
250	12	20.83	5.5215	3.0364
500	20	25.00	6.2146	3.2189
750	27	27.78	6.6201	3.3243
1000	31	32.26	6.9078	3.4738
1250	33	37.88	7.1309	3.6344

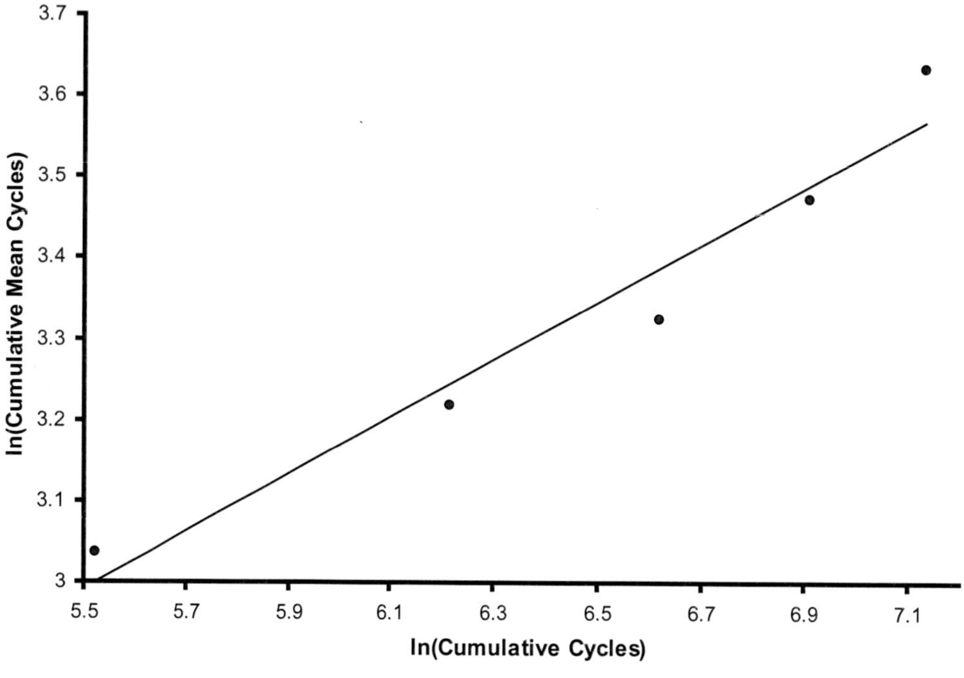

Figure 5.21 *Duane reliability growth plot.*

Solving the above expression for time,

$$75 = \frac{2.864t^{0.353}}{(1 - 0.353)}$$

$$t = 3{,}054 \text{ thousand cycles}$$

AMSAA Model

The AMSAA (Army Material Systems Analysis Activity) model also assumes a non-homogeneous Poisson process. The instantaneous mean time to fail is

$$\theta_i(t) = \frac{\alpha}{\beta}\left(\frac{t}{\alpha}\right)^{1-\beta} \tag{5.30}$$

where

β is the shape parameter

α is the scale parameter

These parameters can be estimated graphically using the expression

$$\ln\left[N(t)\right] = \beta \ln t - \beta \ln \alpha \tag{5.31}$$

where $N(t)$ is the cumulative number of failures at time = t.

By plotting $\ln\left[N(t)\right]$ on the y-axis and $\ln t$ on the x-axis, the slope of the resulting best-fit straight line through the data is the estimated shape parameter β. The scale parameter is determined using the y-intercept

$$\alpha = e^{-\frac{y\text{-intercept}}{\beta}} \tag{5.32}$$

Example 5.18: For the data given in Example 5.17, determine the AMSAA growth model parameters and the number of cycles of testing required to achieve a mean time to fail of 75,000 cycles.

Solution: Table 5.9 contains the data required for plotting.

A plot of the data in Table 5.9 is shown in Figure 5.22.

TABLE 5.9
AMSAA GROWTH PLOT COMPUTATIONS

Cumulative Cycles (Thousands)	Cumulative Failures	Logarithm of Cumulative Cycles (Thousands)	Logarithm of Cumulative Failures
250	12	5.5215	2.4849
500	20	6.2146	2.9957
750	27	6.6201	3.2958
1000	31	6.9078	3.434
1250	33	7.1309	3.4965

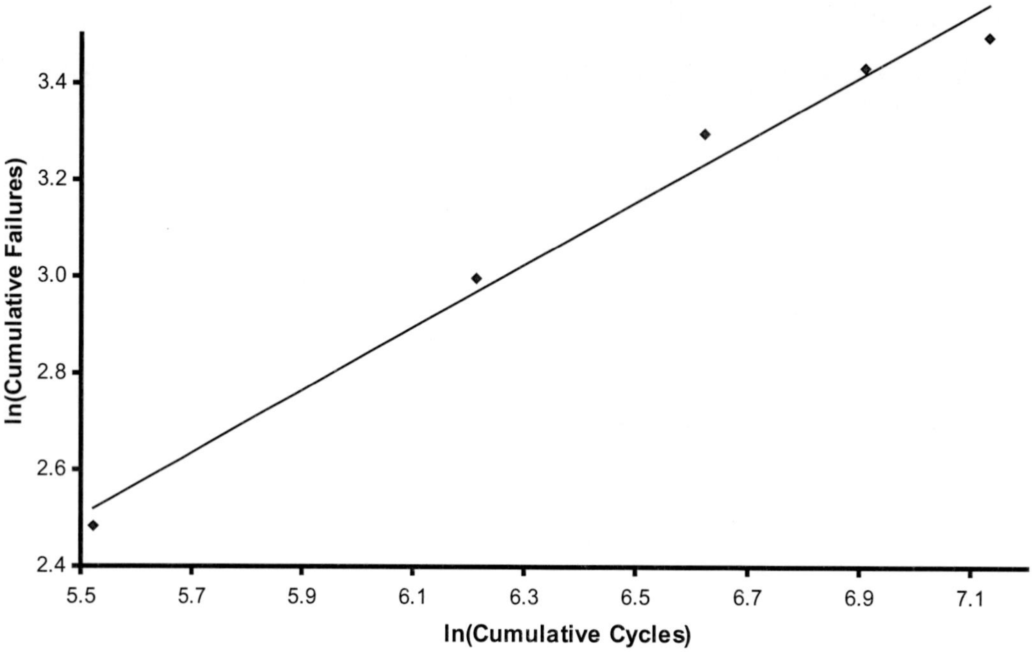

Figure 5.22 *AMSAA reliability growth plot.*

The slope of the best-fit line in Figure 5.22 is 0.647, which is equal to the shape parameter β. The *y*-intercept of the best-fit line in Figure 5.22 is –1.053. The scale parameter is

$$\alpha = e^{-\frac{-1.053}{0.647}} = 5.087 \text{ thousand cycles}$$

The instantaneous mean time to fail is

$$\theta_i(t) = \frac{5.087}{0.647}\left(\frac{t}{5.087}\right)^{1-0.647}$$

Solving for $\theta_i(t) = 75{,}000$ gives

$$t = 3{,}056 \text{ thousand cycles}$$

Summary

There are many approaches to planning an accelerated test. Bayesian and sequential tests can provide efficiency over conventional testing methods. There is no best method for planning an accelerated test. Several types of test plans should be investigated for each situation.

CHAPTER 6

ACCELERATED TESTING MODELS

Accelerated testing models are useful for translating reliability performance across differing stress levels. There are models for single sources of stress and multiple sources of stress. Proper understanding and use of these models can greatly improve the efficiency of accelerated testing.

Linear Acceleration

Although any transformation function could be used to model acceleration, a linear transformation of the time scale is almost always used. Under this assumption, the time to fail under normal operating conditions is

$$t_0 = \varepsilon t_\varepsilon \tag{6.1}$$

where

ε is the acceleration factor

t_ε is the time to fail under increased stress conditions

If $f(t)$ represents the probability density function under accelerated conditions, then the probability density function under normal operating conditions is

$$f_o(t) = \frac{1}{\varepsilon} f\left(\frac{t}{\varepsilon}\right) \tag{6.2}$$

The reliability function under normal operating conditions is

$$R_o(t) = R\left(\frac{t}{\varepsilon}\right) \tag{6.3}$$

The hazard function under normal operating conditions is

$$h_o(t) = \frac{1}{\varepsilon} h\left(\frac{t}{\varepsilon}\right) \tag{6.4}$$

Table 6.1 gives transformed reliability functions for several distributions using an acceleration factor of ε. It can be seen from the reliability functions in this table that there is no acceleration when $\varepsilon = 1$.

TABLE 6.1
RELIABILITY FUNCTIONS WITH AN ACCELERATION FACTOR OF *e*

Distribution	Reliability Function Under Normal Operating Conditions
Weibull	$R(t)=\exp\left[-\left(\frac{t}{\varepsilon\theta}\right)^{\beta}\right]$
Lognormal	$R(t)=1-\Phi\left(\frac{\ln t-\ln\varepsilon-\mu}{\sigma}\right)$
Normal	$R(t)=1-\Phi\left(\frac{t-\varepsilon\mu}{\varepsilon\sigma}\right)$
Exponential	$R(t)=e^{-\frac{\lambda t}{\varepsilon}}$

Example 6.1: When tested with an axial loading of 200 pounds, the time to fail for a component was found to follow a Weibull distribution with a shape parameter of 2.8 and a scale parameter of 44.2. The normal axial loading is 38 pounds, and the acceleration factor between these two load levels is known to be 12. Determine the reliability of the component under normal operating conditions at time = 100.

Solution: The expression for reliability under normal operating conditions is

$$R(t)=\exp\left[-\left(\frac{t}{18(44.2)}\right)^{2.8}\right]$$

At time = 100, the component reliability is

$$R(100)=\exp\left[-\left(\frac{100}{18(44.2)}\right)^{2.8}\right]=0.997$$

Field Correlation

Two conditions are required for an acceleration model to be valid:

1. The failure mode(s) must be the same for all stress levels
2. The shape of the distribution must be the same for all stress levels

One method of comparing the shape of the distributions is to produce probability plots at each stress level. If failure data collected under different levels of stress are plotted on the same probability plot, the lines should be parallel. (The lines will never be exactly parallel because of sampling error.) If parallel lines do not result, either the acceleration is not linear or the chosen probability density function does not adequately model the data.

Another method for verifying that the shape of the distribution has not changed due to increased stress is to compare the confidence intervals for distribution shape parameters. This method is usually applied when the failure data are modeled with the Weibull distribution. If the confidence intervals for the shape parameters do not overlap, then the distribution shapes are not equal. Beware that overlapping confidence intervals do not guarantee that the distribution shapes are equal. With small sample sizes, confidence intervals are large—often too large to discriminate.

Example 6.2: A component was tested at two pressure levels, and Table 6.2 gives the results. The failure modes were identical; therefore, determine if increasing the pressure is a valid method of acceleration.

Solution: Table 6.3 shows the computations required for a Weibull probability plot. Figure 6.1 shows a Weibull probability plot for both sets of data. The lines appear to be parallel; thus, increasing the pressure is a valid method of accelerating the test.

Example 6.3: Determine the acceleration factor for increasing the test pressure from 250 psi to 400 psi for the test described in Example 6.2.

Solution: Using the data from Table 6.3, the scale parameters for the Weibull distribution are

$$250 \text{ psi} - \theta = 352.9$$

$$400 \text{ psi} - \theta = 31.5$$

The acceleration factor for increasing pressure from 250 psi to 400 psi is

$$\varepsilon = \frac{352.9}{31.0} = 11.2$$

In many cases, it is not possible to obtain field data for correlation to laboratory results. When the expected field service of an item is more than 5 years, it is often not practical to wait until field data are available for correlation to laboratory data. Even correlating an accelerated version of a laboratory test to a non-accelerated version is often not feasible due to time constraints. For example, although the equivalent of 10 years' worth of cycles may be applied to an electronic controller in 2 months, with affordable sample sizes (usually fewer than 20), it may take more than a year to obtain enough failures for correlation.

TABLE 6.2
TIME TO FAIL DATA FOR PRESSURE TESTING

Time to Fail (250 psi)	Time to Fail (400 psi)
359	29
207	28
225	28
342	22
282	34
326	44
406	28
483	35
403	21
178	34
326	24
252	28
299	35
351	29
202	15
196	37
459	16
470	33
221	28
340	19
241	38
430	11
	22
	31
	39
	27
	19

A technique for correlation in these situations is to correlate the life remaining in a unit that has been in the field. For example, if a unit has been in the field for 18 months, and 1 hour of accelerated test is the equivalent of 1 month of field use, then it is expected that the unit would survive 18 hours less on the accelerated test than a new unit. One disadvantage of this approach is it requires a large sample size because of the variation in time-to-fail data.

TABLE 6.3
WEIBULL PROBABILITY PLOT COMPUTATIONS

Time to Fail (250 psi)	Time to Fail (400 psi)	Count (250 psi)	Count (400 psi)	$F(t)$ (250 psi)	$F(t)$ (400 psi)	$\ln(t)$ (250 psi)	$\ln(t)$ (400 psi)	$\ln\left[\ln\left(\frac{1}{R}\right)\right]$ (250 psi)	$\ln\left[\ln\left(\frac{1}{R}\right)\right]$ (400 psi)
178	11	1	1	0.0313	0.0255	5.1818	2.3979	–3.4499	–3.6543
196	15	2	2	0.0759	0.0620	5.2781	2.7081	–2.5392	–2.7481
202	16	3	3	0.1205	0.0985	5.3083	2.7726	–2.0523	–2.2659
207	19	4	4	0.1652	0.1350	5.3327	2.9444	–1.7118	–1.9306
221	19	5	5	0.2098	0.1715	5.3982	2.9444	–1.4461	–1.6704
225	21	6	6	0.2545	0.2080	5.4161	3.0445	–1.2254	–1.4557
241	22	7	7	0.2991	0.2445	5.4848	3.0910	–1.0345	–1.2715
252	22	8	8	0.3438	0.2810	5.5294	3.0910	–0.8646	–1.1089
282	24	9	9	0.3884	0.3175	5.6419	3.1781	–0.7100	–0.9623
299	27	10	10	0.4330	0.3540	5.7004	3.2958	–0.5666	–0.8279
326	28	11	11	0.4777	0.3905	5.7869	3.3322	–0.4316	–0.7029
326	28	12	12	0.5223	0.4270	5.7869	3.3322	–0.3027	–0.5854
340	28	13	13	0.5670	0.4635	5.8289	3.3322	–0.1780	–0.4737
342	28	14	14	0.6116	0.5000	5.8348	3.3322	–0.0558	–0.3665
351	28	15	15	0.6563	0.5365	5.8608	3.3322	0.0656	–0.2627
359	29	16	16	0.7009	0.5730	5.8833	3.3673	0.1881	–0.1614
403	29	17	17	0.7455	0.6095	5.9989	3.3673	0.3138	–0.0616
406	31	18	18	0.7902	0.6460	6.0064	3.4340	0.4456	0.0377
430	33	19	19	0.8348	0.6825	6.0638	3.4965	0.5882	0.1373
459	34	20	20	0.8795	0.7190	6.1291	3.5264	0.7494	0.2385
470	34	21	21	0.9241	0.7555	6.1527	3.5264	0.9472	0.3425
483	35	22	22	0.9688	0.7920	6.1800	3.5553	1.2429	0.4511
	35		23		0.8285		3.5553		0.5670
	37		24		0.8650		3.6109		0.6943
	38		25		0.9015		3.6376		0.8404
	39		26		0.9380		3.6636		1.0224
	44		27		0.9745		3.7842		1.2994

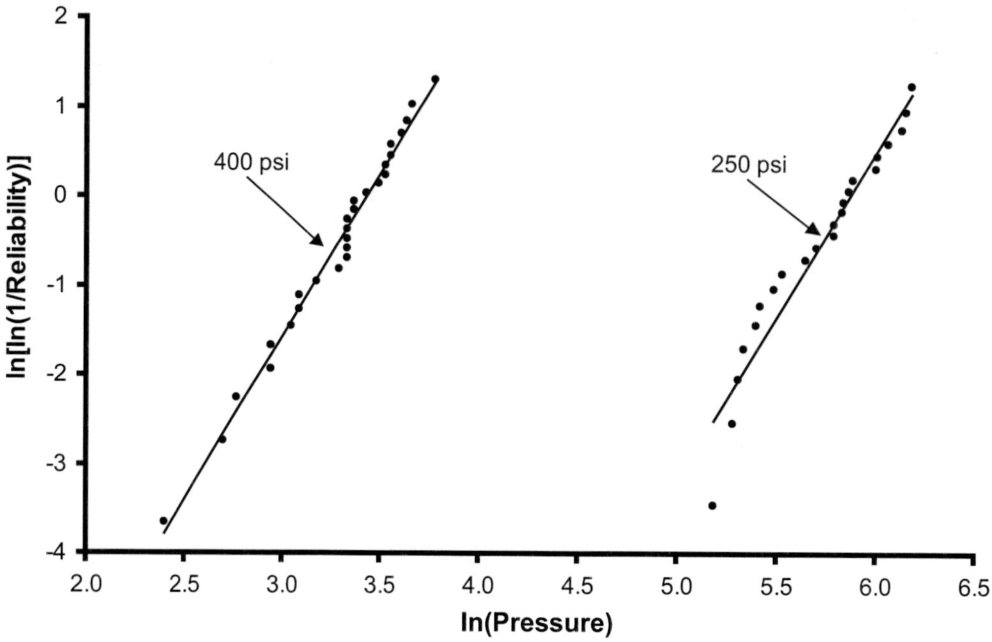

Figure 6.1 *Weibull probability plot for pressure test data.*

Example 6.4: Given the data in Table 6.4, how many miles in the field does 1 hour of testing simulate?

TABLE 6.4
EQUIVALENT LIFE TEST DATA

Miles in Service	Accelerated Testing (Hours to Fail)
4,483	162.9
6,977	253.4
8,392	147.7
8,659	210.0
9,474	266.6
10,243	137.2
10,268	202.9
11,255	219.4
15,206	66.0
15,634	248.1
16,306	116.0
16,681	199.1

TABLE 6.4 ***(Cont.)***

Miles in Service	Accelerated Testing (Hours to Fail)
18,019	192.8
18,974	148.9
20,908	160.8
21,159	242.8
23,002	191.6
23,009	186.9
23,539	135.4
24,678	147.7
25,026	105.4
25,151	205.2
25,533	242.1
26,918	175.3
28,094	171.2
30,634	84.4
31,201	91.3
31,236	129.7
32,502	153.7
33,076	160.4
33,197	125.9
34,618	134.8
35,412	76.9
35,877	116.8
37,978	151.5
38,003	106.6
40,803	145.4
41,187	163.6
41,204	226.3
41,818	117.6
41,951	77.0
42,639	48.6
44,891	110.9
45,929	197.7
46,052	77.9
46,472	136.8

TABLE 6.4 ***(Cont.)***

Miles in Service	Accelerated Testing (Hours to Fail)
49,592	74.4
50,079	85.5
51,168	82.0
51,666	213.8
52,697	67.6

Solution: Figure 6.2 shows a scatter plot of the hours to failure on an accelerated test versus the miles accumulated in field service. The slope of the regression line is –124.7. This means each test hour is the equivalent of 124.7 hours of field service. The poor correlation seen in Figure 6.2 is typical of a remaining life test. For this example, the correlation coefficient is –0.51, which is typical for this type of test.

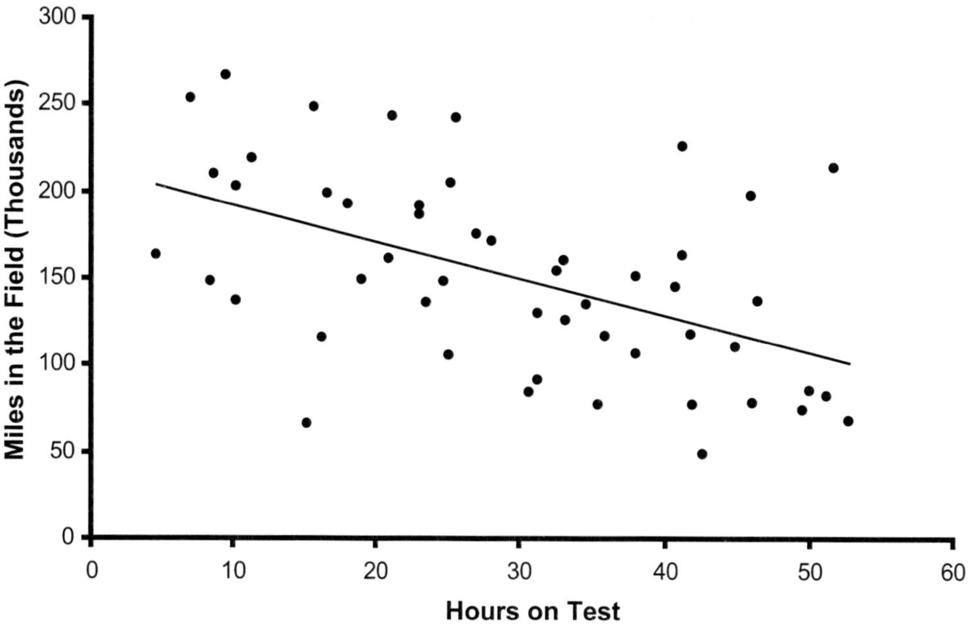

Figure 6.2 Scatter plot of field miles versus laboratory time to fail.

Arrhenius Model

In 1889, the Swedish chemist Svante Arrhenius developed a mathematical model for the influence of temperature on chemical reactions. Because chemical reactions (e.g., lubricant breakdown, corrosion, diffusion of semiconductor material) are responsible for many failures, the Arrhenius equation has been adapted to model the acceleration of reliability testing.

The Arrhenius equation states that the rate of reaction, K, is a function of the absolute temperature T,

$$K = Ae^{-\frac{B}{T}} \tag{6.5}$$

where A and B are constants specific to the reaction conditions.

This equation is often written in the form

$$K = K_0 e^{-\left(\frac{E}{BT}\right)} \tag{6.6}$$

where

K_0 is the reaction rate at a known temperature

E is the activation energy of the reaction

B is Boltzmann's constant

The United States (MIL-HDBK-217), British (HRD5), and French (CNET) governments have developed standards to predict the reliability of electronic equipment using the Arrhenius equation and assuming an exponential time-to-fail distribution. Lall *et al.* (1997), Walker (1998), and O'Connor (2002) have shown that these standards are of little value in predicting the reliability of electronics. Lall *et al.* (1997) state that these standards "have been proven inaccurate, misleading, and damaging to cost-effective and reliable design, manufacture, testing, and support."

This does not mean that the Arrhenius model cannot be used for accelerated testing; simply beware of standards based on steady-state temperatures and exponential time-to-fail distributions. A more general form of the Arrhenius model that allows any desired accelerating stress is

$$t = ke^{\frac{c}{S}} \tag{6.7}$$

where

t is the time at which a specified portion of the population fails

k and c are constants specific to the reaction conditions

S is the level of applied stress

The term S in Eq. 6.7 could be any type of acceleration factor, including:

- Temperature
- Temperature cycling rate
- Vibration level (GRMS)
- Humidity
- Load (pressure or pounds–force)
- Salt concentration
- Voltage cycling rate

By making a logarithmic transformation, Eq. 6.7 can be written as

$$\ln t = \ln k + \frac{c}{S} \tag{6.8}$$

By plotting this expression, the values of the constants in the Arrhenius model can be determined. Kielpinski and Nelson (1975) have developed optimum test plans for the Arrhenius model. These plans estimate the 10th percentile of the time-to-failure distribution at the normal operating stress with the greatest possible large-sample precision. These plans involve testing at only two temperatures, and Meeker and Hahn (1985) caution their use because the plans do not allow any degrees of freedom for assessing goodness of fit. Meeker and Hahn make the following recommendations for developing an accelerated test plan using the Arrhenius model:

- Restrict testing to a range of temperatures over which there is a good chance that the Arrhenius model adequately represents the data.
- Select a second temperature reasonably removed from the highest temperature.
- Select a low temperature that is as close as possible to the design temperature.
- Apportion more of the available test units to the lower levels of stress.

Example 6.5: A component was tested at four levels of axial loading, and Table 6.5 shows the results. (A "+" after a data point indicates the item was removed from testing without failing.) Develop the Arrhenius model for modeling reliability as a function of axial loading.

Solution: The time variable in the Arrhenius model is the time for a specified percentage of the population to fail. A percentage of the population (percentile) could be chosen and determined nonparametrically using the median rank or another nonparametric method, or a distribution could be used to model the data. If the Weibull distribution is used, the scale parameter θ is the point in time where 63.2% of the population fails. Figure 6.3 shows a Weibull probability plot for the hours to fail at the four levels of axial loading.

The Weibull distribution adequately models all data sets, and the four plots (although not perfectly parallel) are close enough to continue with the analysis. Remember that the slope of the probability plot is a sample of the Weibull

TABLE 6.5
ARRHENIUS MODEL SAMPLE DATA

Hours to Fail			
2 Pounds	**5 Pounds**	**10 Pounds**	**20 Pounds**
171.5	39.8	9.1	9.3
178.3	47.1	9.8	9.7
257.2	56.0	11.5	11.4
271.1	57.1	14.4	11.7
333.7	65.9	16.1	13.5
362.9	77.4	20.3	13.8
366.9	79.6	29.4	15.3
400.0 +	85.0	30.8	17.0
400.0 +	88.2	32.6	23.3
400.0 +	89.2	36.1	23.5
400.0 +	100.0 +	36.5	24.7
400.0 +	100.0 +	36.8	24.7
400.0 +	100.0 +	39.3	25.0
400.0 +	100.0 +	40.4	25.1
400.0 +	100.0 +	42.9	26.6
400.0 +	100.0 +	45.2	27.5
400.0 +	100.0 +	47.6	29.4
400.0 +	100.0 +	55.2	30.0

shape parameter, and the estimation method contains error. The plots shown in Figure 6.3 are typical for data from Weibull distributions with equal shape parameters. (The data were generated using a Weibull random number generator with a shape parameter equal to 2.8 for all stress levels.) One important consideration when reviewing the Weibull plots is to ensure that the slope is not changing with the stress level. In Figure 6.3, it can be seen that the slope does change from one data set to another, but there is movement in both directions as the stress is increased. Note that the slope for the data at 2 pounds is nearly identical to the slope for the data at 20 pounds.

For further confirmation of the equality of the four slopes, the results of a maximum likelihood analysis* for the data are shown in Table 6.6.

* For details on performing a maximum likelihood analysis, consult Chapter 4, Parameter Estimation.

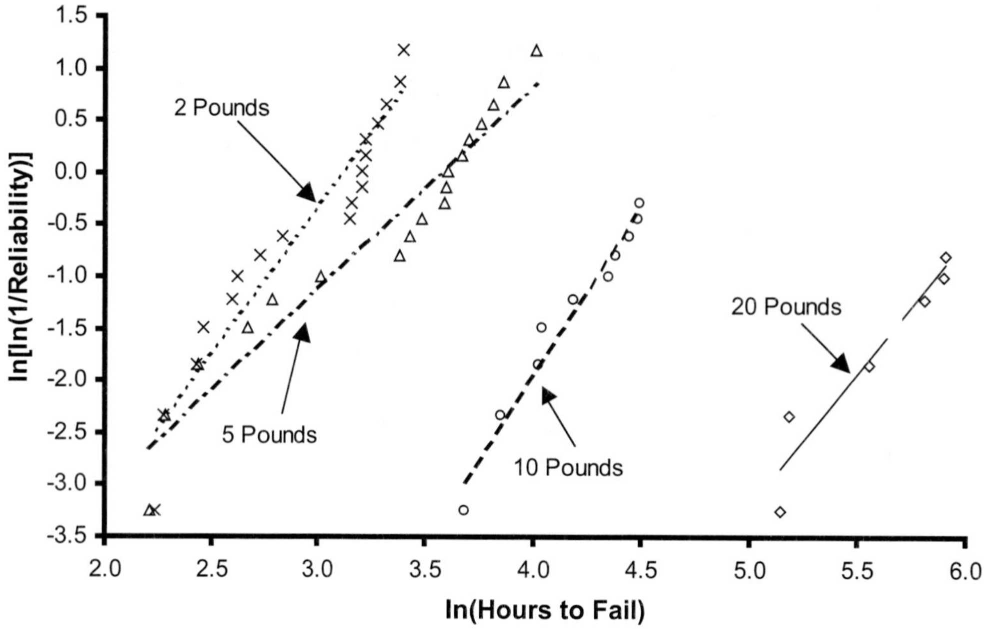

Figure 6.3 Weibull probability plot of axial load data.

TABLE 6.6
WEIBULL MAXIMUM LIKELIHOOD PARAMETER ESTIMATION

Weibull Parameter	2 Pounds	5 Pounds	10 Pounds	20 Pounds
β	2.79	3.02	2.46	3.33
Lower 90% Confidence Limit for β	1.77	2.10	1.91	2.58
Upper 90% Confidence Limit for β	4.39	4.36	3.17	4.29
θ	512.70	105.50	34.75	22.49

As seen in Table 6.6, the confidence intervals for β overlap considerably for all stress levels, confirming that the increase in stress had little or no effect on the distribution shape.

The constants in the Arrhenius model are determined by regressing the inverse of the stress against the logarithm of the scale parameter (θ – the 63.2 percentile), as shown in Figure 6.4.

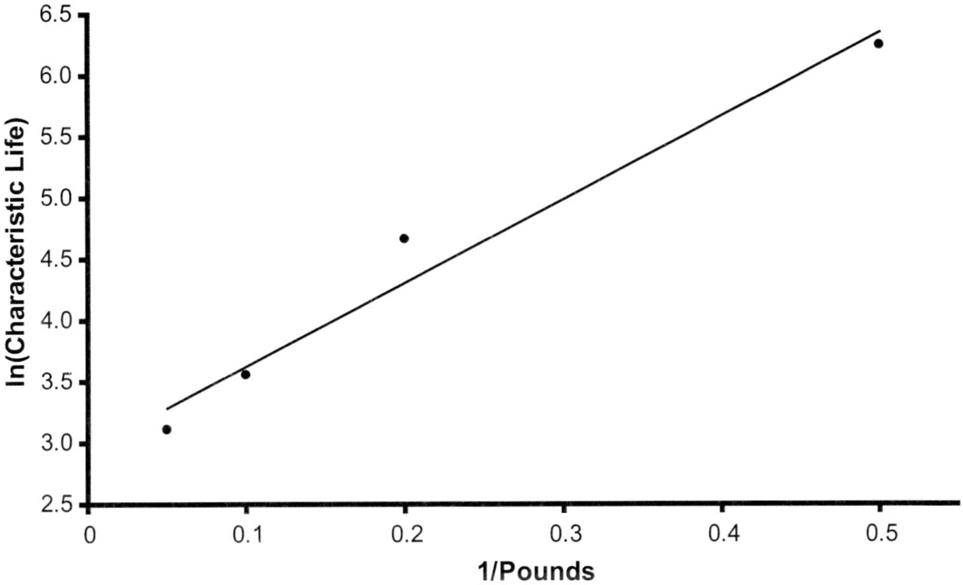

Figure 6.4 *Arrhenius model regression.*

The slope of the regression equation, which provides the estimate for the constant c in the Arrhenius model, is 6.8. The constant k is estimated from the y-intercept, which is 2.942,

$$k = e^{2.942} = 18.9$$

The Arrhenius model for modeling reliability as a function of axial loading is

$$\theta = 18.9e^{\frac{6.8}{\text{pounds}}}$$

Example 6.6: For the model developed in Example 6.5, determine the reliability at 200 hours, given an axial load of 3.5 pounds.

Solution: The estimated scale parameter for an axial load of 3.5 pounds is

$$\hat{\theta} = 18.9e^{\frac{6.8}{\text{pounds}}} = 18.9e^{\frac{6.8}{3.5}} = 131.9 \text{ hours}$$

There were four shape parameters estimated when determining the Arrhenius model, and there was not enough statistical evidence to prove they were not equal. What shape parameter should be used for this analysis? There are several approaches, but a common and simple approach is to use the average of the four estimated shape parameters. The four estimates for the shape parameter were 2.79, 3.02, 2.46, and 3.33. The average of these four values is 2.9. Using a shape parameter of 2.9, the reliability at 200 hours, with an axial load of 3.5 pounds, is

$$R(200) = e^{-\left(\frac{200}{131.9}\right)^{2.9}} = 0.0353$$

Example 6.7: For the data given in Example 6.5, use a nonparametric model to estimate the time that will yield 90% reliability with an axial loading of 4 pounds.

Solution: The cumulative distribution function (i.e., the percentage of the population failing) can be estimated using the median rank,

$$MR = \frac{i - 0.3}{n + 0.4} \tag{6.9}$$

where

i is the order of failure

n is the sample size

The median ranks are the same at all load levels, with the exception of the censored data points. For the second failure, the median rank estimate of $F(t)$ is

$$MR = \frac{2 - 0.3}{18 + 0.4} = 0.092$$

Table 6.7 shows all median rank computations.

Because the requirement is to determine a model for time to fail with 90% reliability, the time for 90% reliability (10% median rank) must be determined. Table 6.8 shows that the time of the second failure occurs at $F(t) = 9.2\%$, and the third failure occurs at $F(t) = 14.7\%$. Table 6.8 shows the linear interpolation to a median rank of 10% for all levels of stress.

The constants in the Arrhenius model are determined by regressing the inverse of the stress against the logarithm of the 10th percentile, as shown in Figure 6.5.

The slope of the regression equation, which provides the estimate for the constant c in the Arrhenius model, is 6.8. The constant k is estimated from the y-intercept, which is 1.99,

$$k = e^{1.99} = 7.3$$

TABLE 6.7
MEDIAN RANK COMPUTATIONS

Order	Hours to Fail				Estimated $F(t)$			
	2 Pounds	5 Pounds	10 Pounds	20 Pounds	2 Pounds	5 Pounds	10 Pounds	20 Pounds
1	171.5	39.8	9.1	9.3	0.038	0.038	0.038	0.038
2	178.3	47.1	9.8	9.7	0.092	0.092	0.092	0.092
3	257.2	56.0	11.5	11.4	0.147	0.147	0.147	0.147
4	271.1	57.1	14.4	11.7	0.201	0.201	0.201	0.201
5	333.7	65.9	16.1	13.5	0.255	0.255	0.255	0.255
6	362.9	77.4	20.3	13.8	0.310	0.310	0.310	0.310
7	366.9	79.6	29.4	15.3	0.364	0.364	0.364	0.364
8		85.0	30.8	17.0		0.418	0.418	0.418
9		88.2	32.6	23.3		0.473	0.473	0.473
10		89.2	36.1	23.5		0.527	0.527	0.527
11			36.5	24.7			0.582	0.582
12			36.8	24.7			0.636	0.636
13			39.3	25.0			0.690	0.690
14			40.4	25.1			0.745	0.745
15			42.9	26.6			0.799	0.799
16			45.2	27.5			0.853	0.853
17			47.6	29.4			0.908	0.908
18			55.2	30.0			0.962	0.962

TABLE 6.8
INTERPOLATION FOR 10% MEDIAN RANK

Percent Fail	Hours to Fail 2 Pounds	5 Pounds	10 Pounds	20 Pounds
9.2%	178.3	47.1	9.8	9.7
14.7%	257.2	56.0	11.5	11.4
10.0%	189.8	48.4	10.0	9.9

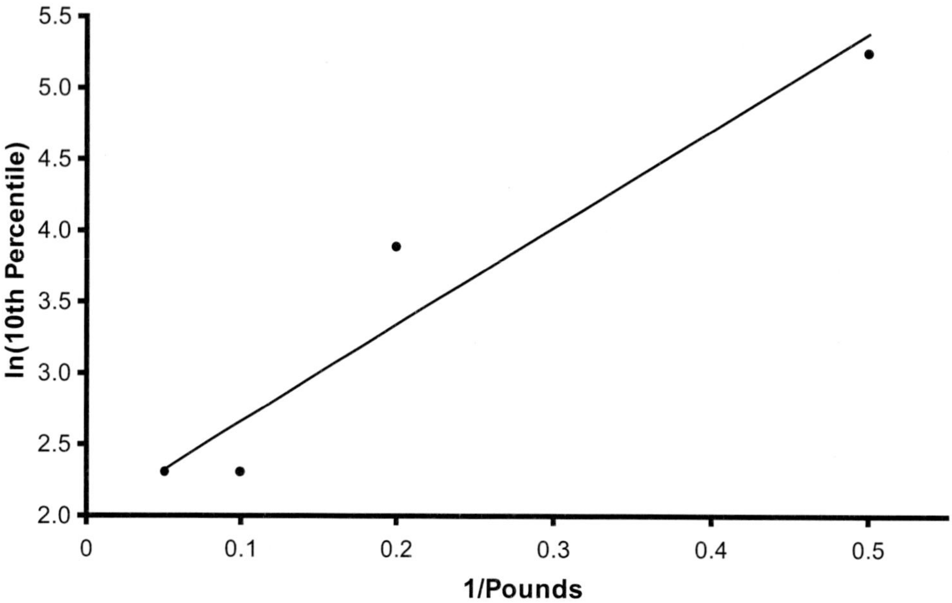

Figure 6.5 *Nonparametric Arrhenius model regression.*

The Arrhenius model for modeling the 10th percentile (i.e., the time at which 90% reliability is achieved) as a function of axial loading is

$$t_{10\%} = 7.3e^{\frac{6.8}{\text{pounds}}}$$

The estimated 10th percentile for an axial load of 4 pounds is

$$t_{10\%} = 7.3e^{\frac{6.8}{4}} = 40.0 \text{ hours}$$

Example 6.8: Determine the Arrhenius model for the characteristic life of the Weibull distribution for the test described in Example 6.2.

Solution: The parameters of the Weibull distribution estimated from probability plotting are shown in Table 6.9.

TABLE 6.9
ESTIMATED WEIBULL PARAMETERS

	250 psi	400 psi
Shape Parameter	3.68	3.53
Scale Parameter	352.9	31.5

With only two data points, regression cannot be used to determine the slope and y-intercept of the best-fit line for the Arrhenius model. The slope of the Arrhenius model is

$$c = \frac{\ln(352.9) - \ln(31.5)}{\left(\frac{1}{250}\right) - \left(\frac{1}{400}\right)} = 1{,}612$$

The y-intercept and the Arrhenius constant k are

$$y\text{-intercept} = \ln(352.9) - \left(\frac{1}{250}\right)(1{,}612) = -0.581$$

$$k = e^{-0.581} = 0.559$$

The Arrhenius model for modeling the characteristic life of the Weibull distribution as a function of pressure is

$$\theta = 0.559e^{\frac{1{,}612}{\text{psi}}}$$

Eyring Model

The Arrhenius model is limited to one type of stress, which is temperature. When more than one type of stress acts on a component, a more complex model is required. Henry Eyring derived a model based on chemical reaction rates and quantum mechanics. If a chemical process (e.g., chemical reaction, diffusion, corrosion, migration) is causing degradation leading to failure, the Eyring model describes how the rate of degradation varies with stress or, equivalently, how time to failure varies with stress. The Eyring model is

$$t = aT^{\alpha}e^{\frac{b}{T}}e^{\left[c+\left(\frac{d}{T}\right)\right]S_1} \tag{6.10}$$

where

t is the time at which a specified portion of the population fails

T is the absolute temperature

S_1 is the level of a second stress factor

α, a, b, c, and d are constants

Although the Eyring model was developed for absolute temperature, the absolute temperature T is sometimes replaced by another stress factor.

The Eyring model can be expanded to include an infinite number of stresses. For every additional stress factor, the model requires two additional constants. The Eyring model with three stress factors is

$$t = aT^{\alpha} e^{\frac{b}{t}} e^{\left[c+\left(\frac{d}{T}\right)\right]S_1} e^{\left[e+\left(\frac{f}{T}\right)\right]S_2} \tag{6.11}$$

The general Eyring model includes terms that have interactions. That is, the effect of changing one source of stress varies, depending on the levels of other stresses. In models with no interaction, acceleration factors can be computed for each stress, and then multiply them together. This would not be true if the physical mechanism required interaction terms. In practice, the Eyring model usually is too complicated to use in its most general form and must be modified or simplified for any particular failure mechanism. With five constants to be estimated for two stress factors in the Eyring model, testing must be conducted under at least five distinct stress combinations. Ideally, more than five testing conditions should be used to demonstrate goodness of fit. Determining the constants is not straightforward and is best done with mathematical software, such as Mathcad®.

A modified version of the Eyring model that eliminates the interaction terms is

$$t = ae^{\frac{b}{T}} S_2^c \tag{6.12}$$

This model originally was developed for accelerating chemical reactions with absolute temperature and voltage, with S_2 being replaced by V (voltage). The interactions are eliminated from the model by setting the constants for the interaction terms, α and d, to 0.

With no interaction terms, only three constants require estimation, and this can be done with multiple linear regression by algebraically manipulating the model into a linear form. The linear form of the Eyring model with no interaction terms is

$$\ln t = \ln a + \frac{b}{T} + c\left(\ln S_2\right) \tag{6.13}$$

The simplified Eyring model can also be extended to include an infinite number of stresses. The general form of this model is

$$\ln t = \ln a + \frac{b}{T} + \sum_{i=1}^{n} \left[c_i \ln S_i \right] \tag{6.14}$$

For example, an Eyring model with four stress sources and no interactions would be

$$\ln t = \ln a + \frac{b}{T} + c_1 \left(\ln S_1 \right) + c_2 \left(\ln S_2 \right) + c_3 \left(\ln S_3 \right) \tag{6.15}$$

Example 6.9: Given the failure time, temperature (degrees Celsius), and salt spray concentration data shown in Table 6.10 (a "+" after a data point indicates the item was removed from testing without failing), develop the Eyring acceleration model for the characteristic life of the Weibull distribution.

TABLE 6.10
EYRING MODEL EXAMPLE DATA

Hours to Fail			
Temp. = 20 Salt = 5%	**Temp. = 20 Salt = 10%**	**Temp. = 50 Salt = 5%**	**Temp. = 50 Salt = 10%**
171.5	39.8	9.1	9.3
178.3	47.1	9.8	9.7
257.2	56.0	11.5	11.4
271.1	57.1	14.4	11.7
333.7	65.9	16.1	13.5
362.9	77.4	20.3	13.8
366.9	79.6	29.4	15.3
400 +	85.0	30.8	17.0
400 +	88.2	32.6	23.3
400 +	89.2	36.1	23.5
400 +	100 +	36.5	24.7
400 +	100 +	36.8	24.7
400 +	100 +	39.3	25.0
400 +	100 +	40.4	25.1
400 +	100 +	42.9	26.6
400 +	100 +	45.2	27.5
400 +	100 +	47.6	29.4
400 +	100 +	55.2	30.0

Solution: When two levels of stress are used for the Eyring model, assuming no interactions, three constants must be determined, which means an experiment with three experimental settings is required. Ideally, more than three experimental trials would be used to allow for the fit of the Eyring model to be checked. Because there are four experimental trials for this example, there is 1 degree of freedom to vary the fit of the model.

The time variable in the Eyring model is the time for a specified percentage of the population to fail. The characteristic life of the Weibull distribution, θ, is the 63.2 percentile. Figure 6.6 shows a Weibull probability plot for the hours to fail at the four levels of salt concentration.

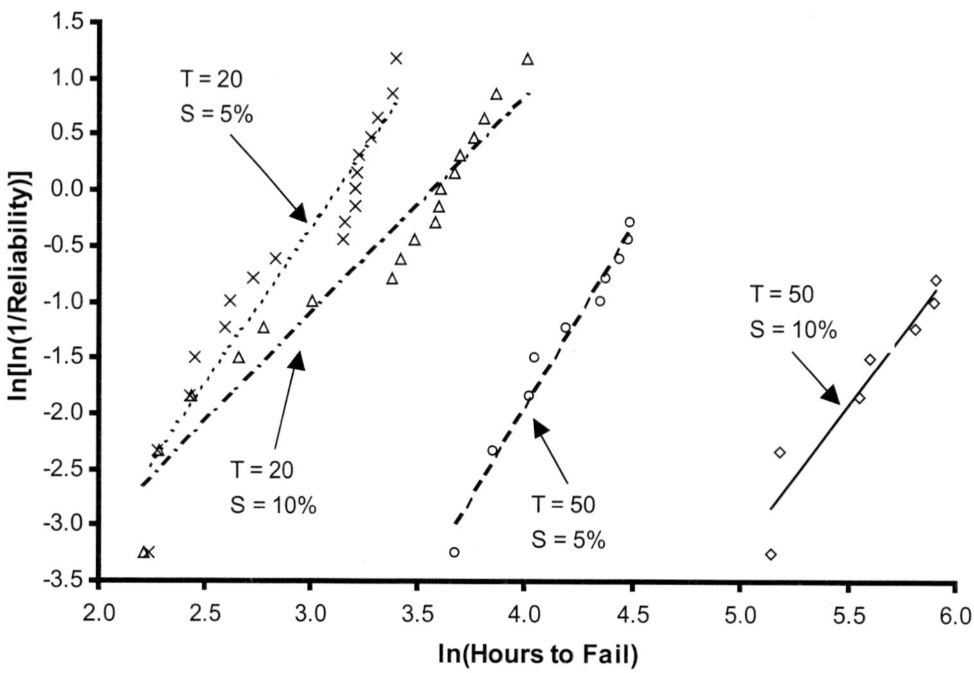

Figure 6.6 *Weibull probability plot of salt concentration data.*

The Weibull distribution adequately models all data sets, and the four plots (although not perfectly parallel) are close enough to continue with the analysis. Remember that the slope of the probability plot is a sample of the Weibull shape parameter, and the estimation method contains error. The plots shown in Figure 6.6 are typical for data from Weibull distributions with equal shape parameters. (These data were generated using a Weibull random number generator with a shape parameter equal to 2.8 for all stress levels.) One important consideration when reviewing the Weibull plots is to ensure that the slope is not changing with the stress level. In Figure 6.6, it can be seen that the slope does change from one data set to another, but there is movement in both directions as the stress is increased.

For further confirmation of the equality of the four slopes, the results of a maximum likelihood analysis* for the data are shown in Table 6.11.

TABLE 6.11
WEIBULL MAXIMUM LIKELIHOOD PARAMETER ESTIMATION

Weibull Parameter	Temp. = 20 Salt = 5%	Temp. = 20 Salt = 10%	Temp. = 50 Salt = 5%	Temp. = 50 Salt = 10%
β	2.79	3.02	2.46	3.33
Lower 90% Confidence Limit for β	1.77	2.10	1.91	2.58
Upper 90% Confidence Limit for β	4.39	4.36	3.17	4.29
θ	512.70	105.50	34.75	22.49

As seen in Table 6.11, the confidence intervals for β overlap considerably for all stress levels, confirming that the increase in stress had little or no effect on the distribution shape.

The original Eyring model uses the absolute temperature ($T + 273$ for Celsius, or $T + 460$ for Fahrenheit). The constants in the Eyring model are determined using multiple linear regression. Table 6.12 shows the data for this regression.

TABLE 6.12
EYRING MODEL DATA FOR MULTIPLE LINEAR REGRESSION

θ	$\ln(\theta)$	Temp. °C	Temp. °K	Salt	$\frac{1}{\text{Temp.}}$	ln Salt
512.7	6.2397	20	293	5%	0.0034	–2.9957
105.5	4.6587	20	293	10%	0.0034	–2.3026
34.8	3.5482	50	323	5%	0.0031	–2.9957
22.5	3.1131	50	323	10%	0.0031	–2.3026

The multiple regression is easily performed using Microsoft® Excel. The following steps describe the process.

1. Select *Data Analysis* from the *Tools* menu.
2. If *Data Analysis* does not appear on the *Tools* menu, select *Add-Ins* from the *Tools Menu*, and check the *Analysis Tool Pack* box, as shown in Figure 6.7.

* For details on performing a maximum likelihood analysis, consult Chapter 4, Parameter Estimation.

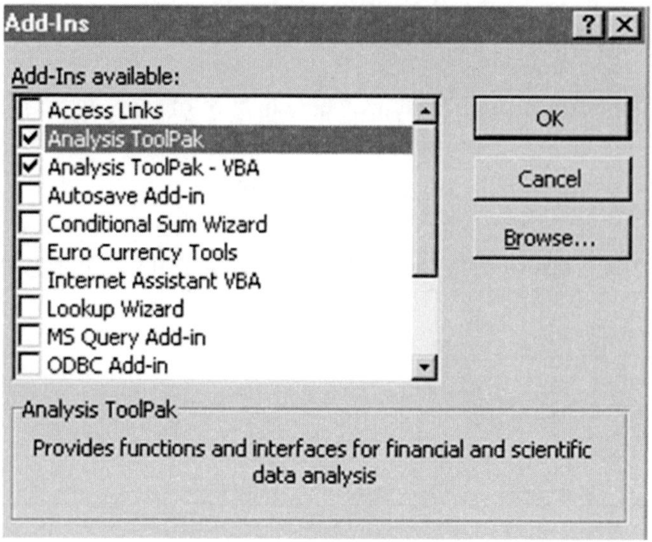

Figure 6.7 *Analysis tool pack.*

3. Select *Regression* from the *Analysis Tools* list.

4. Select the column of data containing the natural logarithm of the characteristic life for the *Input Y Range* field. (Select the column heading along with the data.)

5. Select the two columns of data containing the inverse of absolute temperature and the natural logarithm of the salt concentration for the *Input X Range* field. (Select the column headings along with the data.)

6. Check the *Labels* box.

7. Select a cell for the *Output Range*. This cell will be the upper left corner of the results of the regression.

Figure 6.8 shows steps 4 through 7, and Figure 6.9 shows the results of the regression.

Before determining the constants of the Eyring model, the regression results should be analyzed. The *Adjusted R Square* measures the amount of variation in the data explained by the model. In this case, the regression output indicates that 83.12% of all the variation in the characteristic life is explained by the Eyring model.

The term *Significance F* in the *ANOVA* section is a measure of Type I error for the regression. As a rule of thumb, if the significance is less than 5%, the model adequately represents the data; if the significance is greater than 5%, then the model may not be adequate. In this case, a significance of 23.72% indicates

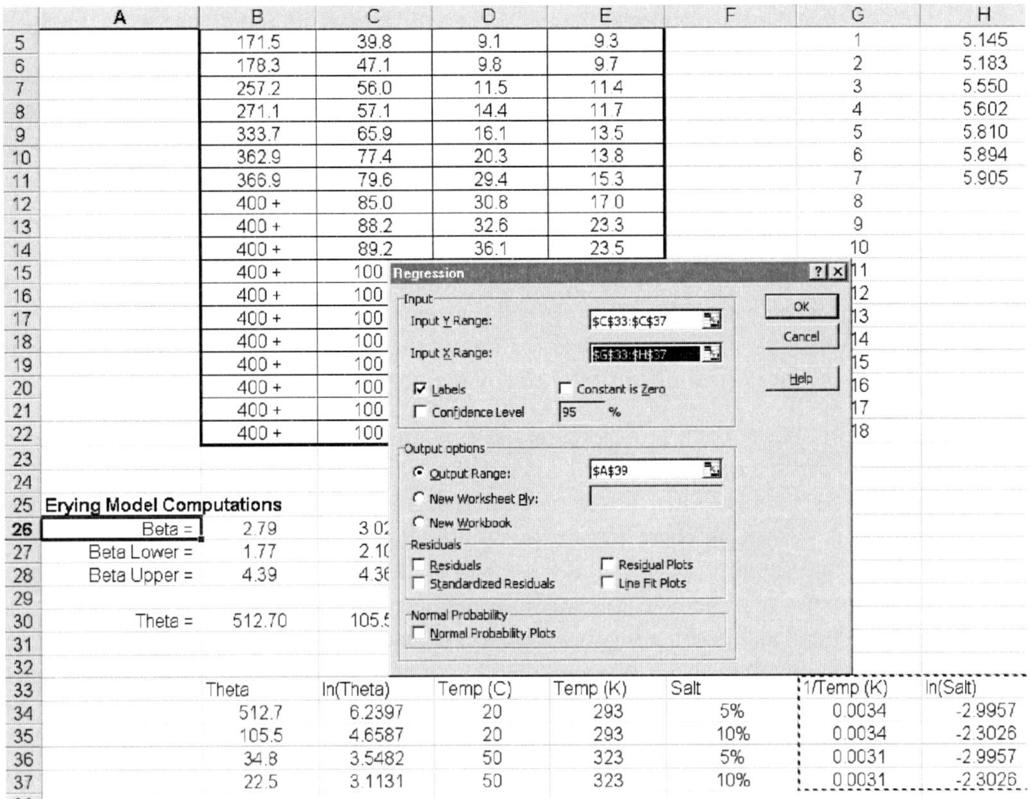

Figure 6.8 *Analysis tool pack multiple linear regression.*

Regression Statistics	
Multiple R	0.9715
R Square	0.9437
Adjusted R Square	0.8312
Standard Error	0.5729
Observations	4

ANOVA

	df	*SS*	*MS*	*F*	*Significance F*
Regression	2	5.5044	2.7522	8.3839	0.2372
Residual	1	0.3283	0.3283		
Total	3	5.8327			

	Coefficients	*Standard Error*	*t Stat*	*P-value*	*Lower 95%*
Intercept	−22.414	6.588	−3.402	0.182	−106.124
1/Temp (K)	7061.833	1909.833	3.698	0.168	−17204.796
ln(Salt)	−1.454	0.827	−1.759	0.329	−11.958

Figure 6.9 *Eyring model regression results.*

that the Eyring model may not be a valid model, although it appears that there is 83.21% of variance in the data set.

The *P-values* for coefficients in the regression provide the Type I error (or significance) for each of the coefficients estimated in the regression model. If the *P-value* is less than 5%, then the slope for the stress factor is not statistically different than 0, which means the stress factor has no proven impact on time to fail. For this example, the Type I error for the regression slope for the inverse of absolute temperature is 16.8%, and the Type I error for the regression slope for the natural logarithm of salt concentration is 32.9%. Based on the Type I error, neither of these factors impacts the time to fail. This is consistent with the lack of significance for the overall regression model.

If neither of the stress factors affects the time to fail, how can the change in the characteristic life be explained? Recall that the characteristic life was 512.7 hours when the temperature was 20°C and the salt concentration was 5%. Likewise, the characteristic life was 22.49 hours when the temperature was 50°C and the salt concentration was 10%. To determine statistical significance, there must be a method for estimating the error of the estimate. The most precise method for estimating this error is to repeat experimental trials. In this example, although each set of experimental conditions (e.g., temperature = 20°C, and salt concentration = 5%) had 18 samples, there was only one result for each set of experimental conditions used in the regression analysis. The regression analysis does use the individual 18 data points. If another 18 samples were tested, and the temperature was 20°C and the salt concentration was 5%, would the resulting characteristic life be 512.7? Without providing repeat trials, the regression assumes that interactions have no statistical significance and uses the degrees of freedom for the interaction to estimate experimental error.

The degrees of freedom for this regression are defined as follows:

Term	Degrees of Freedom
Overall Mean	1
$\frac{1}{\text{Temperature}}$	1
ln(Salt Concentration)	1
Interaction Between $\frac{1}{\text{Temperature}}$ and ln(Salt Concentration)	1

If the interaction has no impact on the time to fail, then the regression is valid; however, if the interaction is present, then the experimental error is overestimated, causing the remaining terms in the model to appear less significant than they really are.

In this example, the conclusion is that the Eyring model with no interaction does not adequately model that data. It appears that an interaction is present. To provide enough degrees of freedom to model the interaction, at least one of the experimental trials should be repeated. If the Eyring model did adequately model the data, the constant a is estimated from the y-intercept,

$$a = e^{y\text{-intercept}} = e^{-22.4} = 1{,}8439\left(10^{-10}\right)$$

The constant b is estimated by the slope for the inverse of absolute temperature, which is 7,061.8. The constant c is estimated by the slope for the natural logarithm of the salt concentration, which is –1.454. The Eyring model for the characteristic life of the Weibull distribution is

$$\theta = 1.8439\left(10^{-10}\right)e^{\frac{7{,}061.8}{T}}S^{-1.454}$$

A second method for checking the fit of the Eyring model to this data is an analysis of residuals, as shown in Table 6.13. There appears to be little or no bias in the residuals, but the error is in the range of 20 to 40% of the predicted value. If this error is sufficient, then the model can be used.

TABLE 6.13
RESIDUALS ANALYSIS OF THE EYRING MODEL

Temperature (°C)	20	20	50	50
Salt Concentration	5%	10%	5%	10%
Actual θ	512.7	105.5	34.75	22.49
Predicted θ	421.94	153.97	44.98	16.41
Residual	–90.76	48.47	10.23	–6.08
Percent Error	–21.5%	31.5%	22.7%	–37.0%

Example 6.10: For the model developed in Example 6.9, determine the reliability with a temperature of 35°C and a salt concentration of 8% at 80 hours in service.

Solution: The estimated scale parameter is

$$\theta = 1.8439\left(10^{-10}\right)e^{\frac{7{,}061.8}{T}}S^{-1.454}$$

$$= 1.8439\left(10^{-10}\right)e^{\frac{7{,}061.8}{(35+273)}}\left(0.08^{-1.454}\right) = 65.9 \text{ hours}$$

Four shape parameters were estimated when determining the Eyring model, and there was not enough statistical evidence to prove they were not equal. What shape parameter should be used for this analysis? There are several approaches, but a common and simple approach is to use the average of the four estimated shape parameters. The four estimates for the shape parameter were 2.79, 3.02, 2.46, and 3.33. The average of these four values is 2.9. Using a shape parameter of 2.9, the reliability at 80 hours, with a temperature of 35°C and a salt concentration of 8%, is

$$R(200) = e^{-\left(\frac{80}{65.9}\right)^{2.9}} = 0.1724$$

Example 6.11: For the data given in Example 6.9, use a nonparametric model to estimate the time that will yield 90% reliability with a temperature of 35°C and a salt concentration of 8%.

Solution: The cumulative distribution function (i.e., the percentage of the population failing) can be estimated using the median rank,

$$MR = \frac{i - 0.3}{n + 0.4} \tag{6.16}$$

where

i is the order of failure

n is the sample size

The median ranks are the same at all load levels, with the exception of the censored data points. For the second failure, the median rank estimate of $F(t)$ is

$$MR = \frac{2 - 0.3}{18 + 0.4} = 0.092$$

Table 6.14 shows all median rank computations.

Because the requirement is to determine a model for time to fail with 90% reliability, the time for 90% reliability (10% median rank) must be determined. Table 6.15 shows that the time of the second failure occurs at $F(t) = 9.2\%$, and the third failure occurs at $F(t) = 14.7\%$. Table 6.15 shows the linear interpolation to a median rank of 10% for all levels of stress.

The original Eyring model uses the absolute temperature ($T + 273$ for Celsius, or $T + 460$ for Fahrenheit). The constants in the Eyring model are determined using multiple linear regression. Table 6.16 shows the data for this regression.

TABLE 6.14
MEDIAN RANK COMPUTATIONS

	Hours to Fail				Estimated $F(t)$			
Order	Temp. = 20 Salt = 5%	Temp. = 20 Salt = 10%	Temp. = 50 Salt = 5%	Temp. = 50 Salt = 10%	Temp. = 20 Salt = 5%	Temp. = 20 Salt = 10%	Temp. = 50 Salt = 5%	Temp. = 50 Salt = 10%
1	171.5	39.8	9.1	9.3	0.038	0.038	0.038	0.038
2	178.3	47.1	9.8	9.7	0.092	0.092	0.092	0.092
3	257.2	56.0	11.5	11.4	0.147	0.147	0.147	0.147
4	271.1	57.1	14.4	11.7	0.201	0.201	0.201	0.201
5	333.7	65.9	16.1	13.5	0.255	0.255	0.255	0.255
6	362.9	77.4	20.3	13.8	0.310	0.310	0.310	0.310
7	366.9	79.6	29.4	15.3	0.364	0.364	0.364	0.364
8		85.0	30.8	17.0		0.418	0.418	0.418
9		88.2	32.6	23.3		0.473	0.473	0.473
10		89.2	36.1	23.5		0.527	0.527	0.527
11			36.5	24.7			0.582	0.582
12			36.8	24.7			0.636	0.636
13			39.3	25.0			0.690	0.690
14			40.4	25.1			0.745	0.745
15			42.9	26.6			0.799	0.799
16			45.2	27.5			0.853	0.853
17			47.6	29.4			0.908	0.908
18			55.2	30.0			0.962	0.962

TABLE 6.15
INTERPOLATION FOR 10% MEDIAN RANK

	Hours to Fail			
Percent Fail	**Temp. = 20 Salt = 5%**	**Temp. = 20 Salt = 10%**	**Temp. = 50 Salt = 5%**	**Temp. = 50 Salt = 10%**
9.2%	178.3	47.1	9.8	9.7
14.7%	257.2	56.0	11.5	11.4
10.0%	189.8	48.4	10.0	9.9

TABLE 6.16
EYRING MODEL DATA FOR MULTIPLE LINEAR REGRESSION

10th Percentile	ln(10th Percentile)	Temp. °C	Temp. °K	Salt	$\frac{1}{\text{Temp.}}$	ln(Salt)
189.8	5.2458	20	293	5%	0.0034	–2.9957
48.4	3.8794	20	293	10%	0.0034	–2.3026
10.0	2.3073	50	323	5%	0.0031	–2.9957
9.9	2.2973	50	323	10%	0.0031	–2.3026

The constant a is estimated from the y-intercept,

$$a = e^{y\text{-intercept}} = e^{23.68} = 5.17522\left(10^{11}\right)$$

The constant b is estimated by the slope for the inverse of absolute temperature, which is 7,534.3. The constant c is estimated by the slope for the natural logarithm of the salt concentration, which is –0.993. The Eyring model for the 10th percentile (90% reliability) is

$$t = 5.17522\left(10^{-11}\right)e^{\frac{7{,}534.3}{T}}S^{-0.993}$$

For 35°C and 8% salt concentration, the time at which 90% reliability is achieved is

$$t = 5.17522\left(10^{-11}\right)e^{\frac{7{,}534.3}{T}}S^{-0.993}$$

$$= 5.17522\left(10^{-11}\right)e^{\frac{7{,}534.3}{(35+273)}}\left(0.08^{-0.993}\right) = 26.7 \text{ hours}$$

Voltage Models

There are many empirically derived acceleration models for voltage, humidity, vibration, and other acceleration factors. The following empirical relationship has been shown to hold true for accelerating failure by changing voltage levels:

$$\frac{t}{t_\varepsilon} = \left(\frac{V_\varepsilon}{V}\right)^3 \tag{6.17}$$

where

t is the time at which a specified portion of the population fails under normal operating conditions

t_ε is the time at which a specified portion of the population fails under accelerated operating conditions

V is the normal voltage

V_ε is the accelerated level of voltage

The inverse voltage power model, shown in Eq. 6.18, is a simplified Eyring model that has been shown to be useful in predicting the life of capacitors:

$$t = aV^{-b} \tag{6.18}$$

where

t is the time at which a specified portion of the population fails

V is voltage

a and b are constants to be determined

Lifetime as a function of voltage has been shown to follow the exponential model shown in Eq. 6.19:

$$t = ae^{bV} \tag{6.19}$$

where

t is the time at which a specified portion of the population fails

V is voltage

a and b are constants to be determined

A modified form of the Eyring model has been proven useful for modeling the acceleration of electromigration failures in semiconductors. The ionic movement is accelerated by temperatures and current density. The modified Eyring model for semiconductor electromigration is

$$t = aJ^{-b}e^{\frac{c}{T}} \tag{6.20}$$

where

t is the time at which a specified portion of the population fails

J is the current density

T is the absolute temperature

a, b, and c are constants to be determined

Mechanical Crack Growth

The Coffin-Mason model was developed to model material cracking, fatigue, and deformation as a function of temperature cycles. This number of cycles to fail is modeled by the expression

$$t = aF^{b}T^{c}$$

where

t is the number of cycles at which a specified portion of the population fails

F is the cycling frequency

T is the temperature range during a cycle

a, b, and c are constants to be determined

Example 6.12: For the data given in Table 6.17, determine the parameters of the Coffin-Mason acceleration model, and determine the number of cycles that gives 90% reliability for 2 temperature cycles per hour, with a minimum temperature of –20°C and a maximum temperature of 100°C. Note that the 10th percentile represents the time at which 10% of the population fails.

Solution: A linear version of the Coffin-Mason model is

$$\ln t = \ln a + b \ln F + c \ln T$$

The coefficients of this equation can be found with multiple linear regression, using the natural logarithm of the 10th percentile as the independent variable and the natural logarithm of F and the natural logarithm of T as the dependent

TABLE 6.17
COFFIN-MASON EXAMPLE DATA

F	*T*	10th Percentile
1	80	487.1
5	80	257.3
1	150	301.9
5	150	82.4

variables. (Example 6.9 shows the procedure for using Microsoft Excel for multiple linear regression.)

Using multiple linear regression,

$$a = 161{,}078$$

$$b = -0.602$$

$$c = -1.286$$

The number of cycles that gives 90% reliability is

$$t = 161{,}078F^{-0.602}T^{-1.286}$$

The number of cycles that gives 90% reliability for 2 temperature cycles per hour, with a minimum temperature of –20°C and a maximum temperature of 100°C, is

$$t = 161{,}078\left(2^{-0.602}\right)\left(120^{-1.286}\right) = 224.7$$

Degradation Testing

Modeling performance degradation can dramatically reduce testing duration and sample sizes. Two features of degradation testing allow efficiency in test duration and sample size:

1. The time to fail is predicted with a model of how performance degrades over time
2. The statistical methods use continuous data from the performance parameters that degrade, rather than binomial, pass-fail statistics

The ability to predict performance allows a test to be abandoned early if the model predicts that the reliability requirements will not be achieved, saving testing resources. Testing also may be stopped early if the model predicts a successful test, but extrapolating for a successful test is much riskier than extrapolating to a prediction of failure.

The disadvantage of degradation testing is that a model is required for the degradation. The acceleration models described in previous sections can be used as degradation models. Models for specific phenomena are also available.

Degradation testing is commonly used for metal creep, crack initiation, crack propagation, tensile strength, flexural strength, corrosion, rust, elongation, breakdown voltage, resistance, and many others.

Consider a simple case of linear degradation, such as material wear. (Examples of this are brake pads and tire tread.) A component fails as material is removed by use. When a component is new, the material thickness is 40 millimeters. The component is considered to be in a failed state when the material thickness is less than 8 millimeters. Table 6.18 gives test data for four components, and Figure 6.10 shows a plot of these data.

TABLE 6.18
LINEAR DEGRADATION EXAMPLE DATA

Test Duration (Hours)	Material Thickness (Millimeters)			
	Sample A	Sample B	Sample C	Sample D
10	36.8	36.5	36.9	37.1
20	33.4	35.3	36.0	33.8
30	29.7	32.2	33.6	31.7
40	25.8	28.8	30.3	29.3
50	25.2	27.1	27.4	26.8

From Figure 6.10, it can be seen that assumption of linear degradation is valid through 50 hours of use. Without prior experience, extrapolating beyond 50 hours could cause errors. Material thickness as a function of test duration is modeled with the expression

$$y = a + bx + e$$

where

y is the material thickness

x is the test duration

e is the error in the model

The parameters in this model can be estimated using least squares regression:

$$b = \frac{n\sum_{i=1}^{n} x_i y_i - \left(\sum_{i=1}^{n} x_i\right)\left(\sum_{i=1}^{n} y_i\right)}{n\sum_{i=1}^{n} x_i^2 - \left(\sum_{i=1}^{n} x_i\right)} \tag{6.21}$$

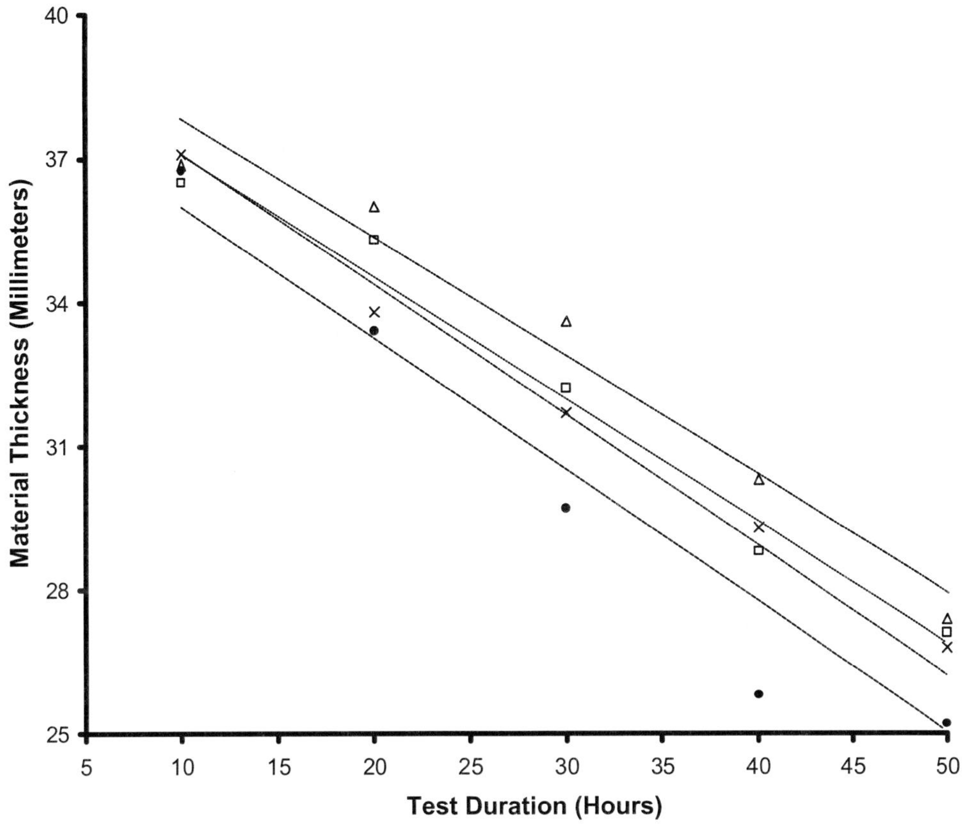

Figure 6.10 *Linear degradation example data.*

$$a = \frac{\sum_{i=1}^{n} y_i - b\sum_{i=1}^{n} x_i}{n} \tag{6.22}$$

where n is the sample size.

A prediction interval that contains a single response with a confidence of $(1 - \alpha)$ is

$$\hat{y}_0 \pm t_{\frac{\alpha}{2}} s\sqrt{1 + \frac{1}{n} + \frac{(x_0 - \bar{x})^2}{S_{xx}}} \tag{6.23}$$

where t is the critical value of the t-distribution with $n - 2$ degrees of freedom, and

$$s = \sqrt{\frac{S_{yy} - bS_{xy}}{n - 2}} \tag{6.24}$$

$$S_{xx}=\sum_{i=1}^{n}x_i^2-\frac{\left(\sum_{i=1}^{n}x_i\right)^2}{n} \tag{6.25}$$

$$S_{yy}=\sum_{i=1}^{n}y_i^2-\frac{\left(\sum_{i=1}^{n}y_i\right)^2}{n} \tag{6.26}$$

$$S_{xy}=\sum_{i=1}^{n}x_iy_i-\frac{\left(\sum_{i=1}^{n}x_i\right)\left(\sum_{i=1}^{n}y_i\right)}{n} \tag{6.27}$$

A confidence interval that contains the mean response with a confidence of $(1-\alpha)$ is

$$\hat{y}_0 \pm t_{\frac{\alpha}{2}} S \sqrt{\frac{1}{n}+\frac{(x_0-\bar{x})^2}{S_{xx}}} \tag{6.28}$$

Solving Eqs. 6.21 and 6.22, the predicted material thickness is

$$y = 39.63 - 0.265x$$

Recall that the component is considered to be in a failed state when the material thickness is less than 8 millimeters. If the component is required to function for 200 hours, the expected material thickness would be

$$y = 39.63 - 0.265(200) = -13.3$$

Thus, there is no need to continue this test for 200 hours. Failure can be predicted at 50 hours.

Now consider the same component in the preceding example, with a change in requirements. When a component is new, the material thickness is 40 millimeters. The component is considered to be in a failed state when the material thickness is less than 8 millimeters, and the required life of the component is 105 hours. The expected material thickness after 105 hours of use would be

$$y = 39.63 - 0.265(105) = 11.8$$

The expected material thickness meets the requirement. Confidence limits and prediction limits for the expected material thickness can be computed using the method of maximum likelihood (Young, 1998) or by using the error of the regression model. Using the regression model, a 90% confidence interval for the mean material thickness after 105 hours of operation is

$$9.2 \le y \le 14.5$$

A 90% prediction interval for the material thickness after 105 hours of operation is

$$8.4 \le y \le 15.2$$

After 105 hours of operation, there is a 95% chance (one side of the 90% confidence interval) that the mean value of the material thickness is greater than 9.2, and it is expected that 90% of the population will have a material thickness greater than 8.4. Testing could be suspended at this point because predictions indicate success. Caution should be taken when accelerated degradation testing is used to cease testing early in cases of success. Extrapolating accelerated degradation test models could cause invalid test results.

Example 6.13: Table 6.19 gives degradation data for 5 prototype switches. The designed life of the switch is the equivalent of 155 hours of testing.

a. Projecting to 155 hours of testing, 90% of the population will have a resistance less than what value?

b. If the maximum design resistance is 550, what decision should be made after 70 hours of testing?

TABLE 6.19
SWITCH DEGRADATION DATA

Test Duration	Resistance				
(Hours)	Switch A	Switch B	Switch C	Switch D	Switch E
10	240.7	255.5	293.3	268.7	295.5
20	316.2	297.9	316.1	305.0	348.1
30	380.7	392.4	343.7	338.2	362.0
40	409.1	436.0	397.5	359.5	400.9
50	419.2	459.8	429.2	442.5	445.3
60	455.8	472.8	444.0	464.6	455.3
70	475.8	479.8	449.3	468.9	470.2

Solution: Figures 6.11 and 6.12 show linear and logarithmic models fitted to the switch degradation data in Table 6.19.

As seen from Figures 6.11 and 6.12, the logarithmic model fits the degradation data better than the linear model. Linear regression can be used to project degradation by transforming the logarithm of x in place of x. (This example is contained in Microsoft Excel on the accompanying CD.) Using the logarithmic model, the expected switch resistance after 155 hours is 553.1, and the upper 95% prediction limit for resistance is 600. Given a design resistance requirement of 550, testing should be stopped at this point; this switch does not meet reliability requirements.

Qualitative Tests

Because of the difficulty of predicting field performance from accelerated tests, several types of tests are designed to reveal failure modes and performance limits with little or no regard for field correlation. These types of tests have three purposes:

a. To define the operating envelope

b. To discover failure modes that may or may not be addressed

c. To compare

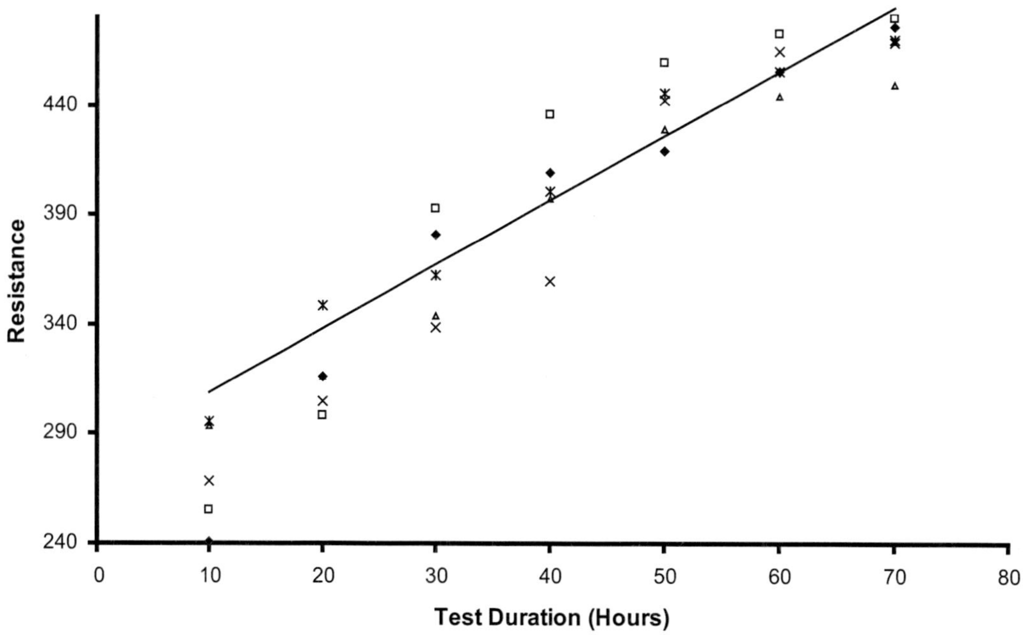

Figure 6.11 *Linear model of switch degradation data.*

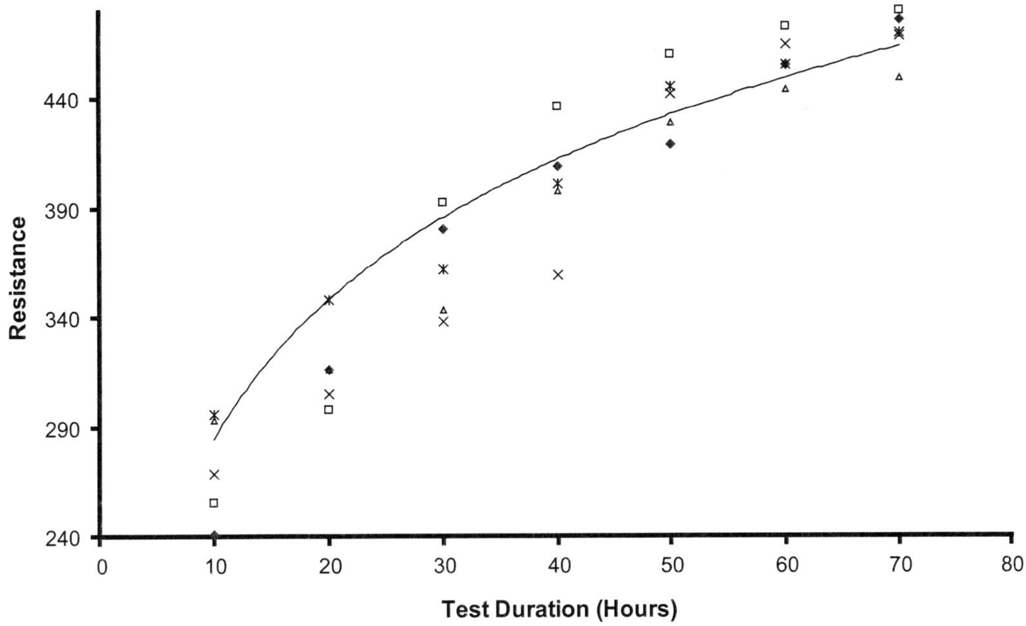

Figure 6.12 *Logarithmic model of switch degradation data.*

Step-Stress Testing

The purpose of step-stress testing is to define an operating envelope. A component is subjected to successively higher levels of stress until the component fails or the limits of the test equipment are reached. Figure 6.13 shows a typical step-stress test plan.

Step-stress testing may involve one or more levels of stress. In some cases, all but one type of stress are held constant; in some cases, the level changes for multiple levels of stress. For example, the temperature may be held constant while the vibration level is increased, or both the temperature and vibration may be increased during a single step-stress test.

An area of concern with step-stress testing is that the failures are a result of accumulated damage rather than the latest level of stress. For example, consider a component that fails after 55 minutes of operation due to a design flaw. This component is exposed to a step-stress test to determine the maximum operating temperature. Table 6.20 gives the test plan. The step-stress plan would show failures at 60°C and incorrectly assume the maximum operating temperature is 50°C.

This problem can be avoided by testing multiple samples at different beginning levels of stress. For example, a second sample could start testing at 30°C, and a third sample could begin testing at 40°C.

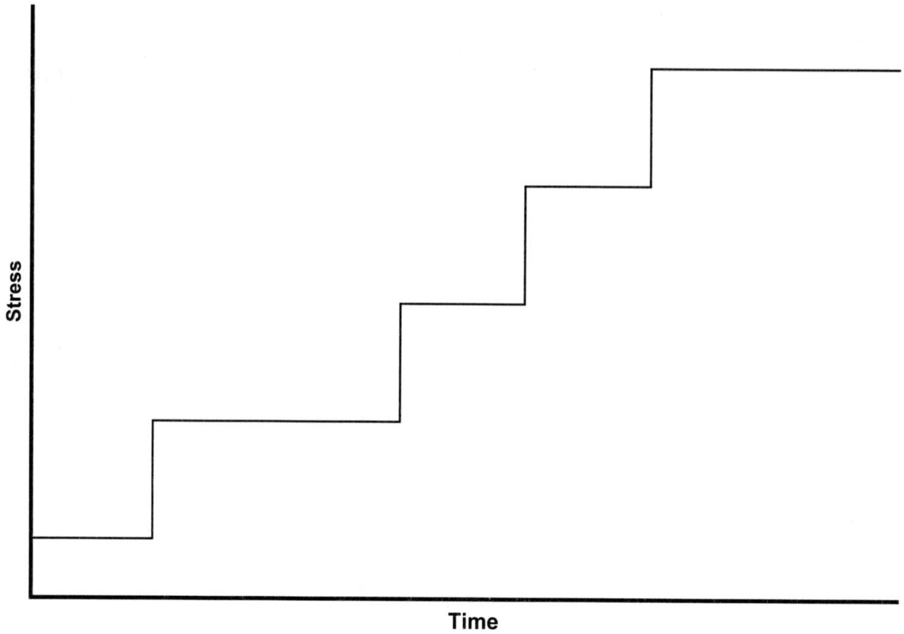

Figure 6.13 Typical step-stress test.

TABLE 6.20
STEP-STRESS TEMPERATURE TEST PLAN

Time	Test Temperature
8:00 am	20°C
8:05 am	25°C
8:10 am	30°C
8:15 am	35°C
8:20 am	40°C
8:30 am	45°C
8:40 am	50°C
8:50 am	60°C
9:00 am	70°C
9:10 am	80°C
9:20 am	90°C
9:30 am	100°C
9:40 am	110°C
9:50 am	120°C
10:00 am	End Test

Elephant Tests

Elephant testing is named because the test "steps on the product with the force of an elephant." Elephant tests do not attempt to correlate to field life; they merely try to precipitate failures. When failures are found, those failures may be

- Ignored
- Corrected to improve reliability
- Used for comparison to another supplier, another design, or another production lot

Elephant tests may contain one or more types of stress, and each type of stress may be at one or more levels. Products are commonly subjected to more than one elephant test, with each test designed for a particular type of failure mode.

*HALT and HASS**

Highly accelerated life testing (HALT) is designed to discover failure modes. The temperature and vibration levels used during a HALT test may purposely be different than actual operating conditions. During the HALT process, a product is subjected to increasing stress levels of temperature and vibration (independently and in combination), rapid thermal transitions, and other stresses specifically related to the operation of the product.

The information goals of HALT are as follows:

- To determine multiple failure modes and root causes
- To determine functional operating limits
- To determine functional destruct limits

HALT was developed specifically for solid-state electronics but can be applied to any product, with the correct equipment. In general, the information goal of HALT is most effectively met by testing at the lowest possible subassembly. Consider a balance between a "functional assembly" and determining the technological limit of a product feature. Consider removing structure that dampens vibration or blocks air flow, reducing the applied stresses to subassemblies. Consider that loss of critical circuitry or connections may affect the subassembly tested.

A complaint with HALT is the inability to reproduce failure modes. This sometimes occurs because of the random nature of HALT tests. Especially with the use of air hammers, vibration is random and uncontrolled across a frequency range. If a test is repeated, the vibration exposure will likely be different.

The results of a HALT test often are used to design a highly accelerated stress screen (HASS). A typical HASS applies all stresses simultaneously. During vibration, the temperature is continuously ramped between brief dwells at extremes. Some typical HASS properties are as follows:

* HALT and HASS material is used with permission of Entela, Inc., www.entela.com.

- Includes all critical product stresses, not only temperature and vibration. For example, a HASS test may include power cycling, voltage spikes, and other environmental factors.
- A HASS test may screen and evaluate the product beyond operating limits and near destruct limits.
- A HASS test precipitates failures in product due to latent defects.

Summary

There are many types of accelerated tests. Some of the tests are quantitative, and some are qualitative. It is recommended to consider the limitations of the quantitative models when making business decisions. For example, how well can a test with a duration of weeks (or less) mimic the field usage of a product for 10 years? Also, the statistical confidence computed from quantitative tests requires a random sample representative of the population that will be placed in the field. These tests are usually prototypes and do not account for all sources of variation, such as:

- Changing operators
- Multiple batches of raw materials
- Changing suppliers
- Equipment maintenance
- Shift changes

Before deciding on a testing plan, review the limits of each test when assessing the alternatives.

CHAPTER 7

ENVIRONMENTAL STRESS SCREENING

Accelerated testing models are often used to create screening or burn-in programs for use in production. This chapter presents the details of environmental stress screening (ESS) and a model to financially optimize the return on investment.

Stress Screening Theory

Electronic hardware product reliability requires both an understanding of why products fail and the advantages of finding, forcing, and correcting these failures. Of critical importance is being able to address the types of product failure modes that may not be discovered through inspection or testing, such as the following:

- Parameter drift
- Printed circuit board shorts and opens
- Incorrect installation of a part
- Installation of the wrong part
- Contaminated part
- Hermetic seal failure
- Foreign material contamination
- Cold solder joints
- Defective parts

The Product Reliability Equation

In efforts to qualify components, assemblies, and finished systems, electronics manufacturers have placed a concerted emphasis on reliability, to the extent that reliability is now generally considered a factor in product performance.

The reason for this is that even well-designed products, manufactured through an accepted or mature process environment, can fail. The "how" and "why" of many of these failures can be traced to a number of factors, although most fall within three categories:

1. Defective components
2. Workmanship error
3. Process stress

These are typically defined as latent defects, which are flaws that are not readily apparent—nor identifiable through conventional testing—but that, with time, can precipitate product failure.

Cumulative failure data indicate that latent defects generally lead to failures early in the life of a product. Once a product passes through this stage, it is usually reliable, as shown in a typical life cycle curve in Figure 7.1, often referred to as a bathtub curve, or in the tri-modal distribution of failure density plotted over the life of a product.

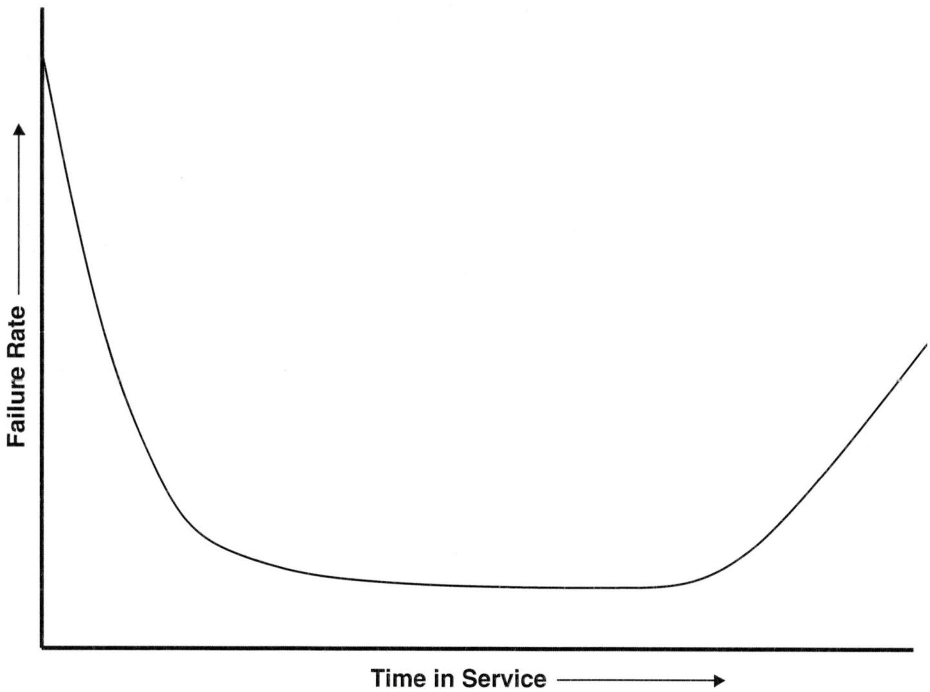

Figure 7.1 *Bathtub curve.*

From this basic model, life cycle patterns can be plotted, as shown in Figure 7.2. The first break, where most failures can be traced, is referred to as infancy. The center section of the curve is the useful life of a product.

The final upsweep represents the normal wearout period of a product. The critical factor to controlling product reliability is to ensure that products have successfully passed through infancy prior to delivery to a customer. Put another way, it requires establishing a means of precipitating infancy failures in-house.

Customer satisfaction is only one reason why an intensified focus on reliability is necessary. Others include the following:

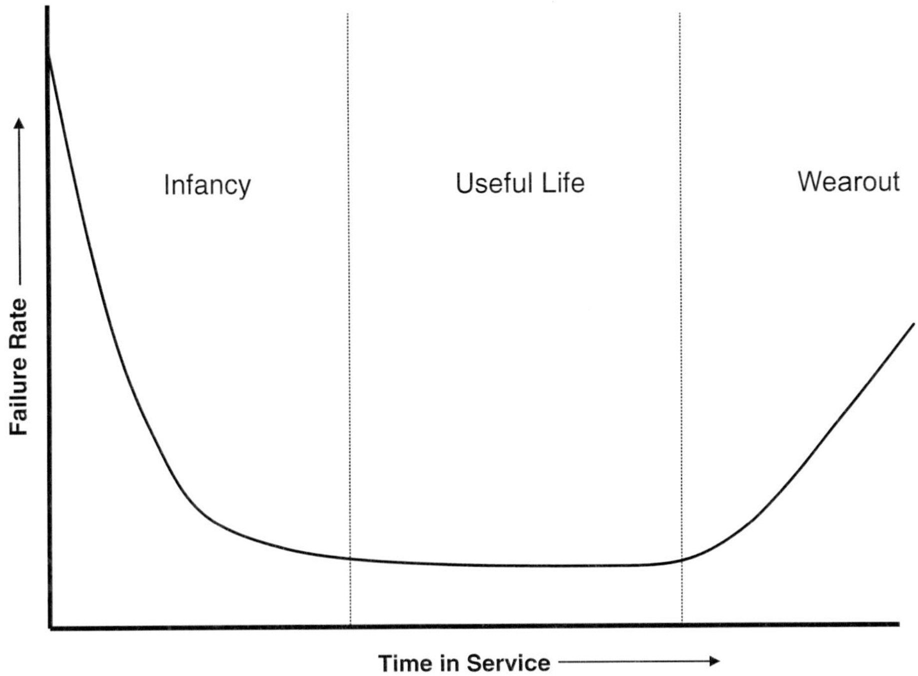

Figure 7.2 Life cycle failure patterns.

- **Mandated Qualification**—Both the consumer and military industries have stepped up emphasis on reliability certification. Some even include a penalty clause for failure to prove reliability as part of a contract.
- **Short Market Life Cycles**—Rapidly changing technologies and increased competition diminish the amount of attention or interest a product can generate. A defective introductory product can virtually kill a company's opportunity in that market.
- **High Rework Costs**—Repairing or replacing components at later phases of production is both time and cost intensive.
- **High Warranty Costs**—The expense of having to disassemble a product in the field, or administer a product recall, can have both a hidden and needless impact on unit cost.
- **Lost Market Share**—Continued customer dissatisfaction can have a long-term impact that not only affects profitability but may require increased marketing and public relations expenditure to counter.

Resolving these concerns has led to a reexamination of qualification theory. More conventional forms of testing, in which failure-free product operation is desired, are being augmented by ESS, in which failures are both expected and necessary to assure parts and workmanship reliability.

What Is ESS?

By basic definition, environmental stress screening (ESS) consists of five screening products, ideally at the most cost-effective point of assembly, to expose defects that cannot be detected by visual inspection or electrical testing. These defects typically are related to defective parts, workmanship, or process and are major contributors to early product field failure.

Through an ESS program, 100% of a group of products is subjected to an environmental stimulus, or a set of stimuli, for a predetermined time for the purpose of forcing failures to occur in-house.

Critical to an understanding of ESS is that, within the program, failures are expected, normal, and unavoidable. In this sense, ESS is radically different from conventional certification testing, which requires failure-free operation as proof of reliability. It is equally important to realize that ESS is not simply an amplified certification test forcing greater, and possibly damaging, stress on the product. In fact, it is not a test at all, but a program used to upgrade product reliability in ways testing cannot.

Toward this end, two important factors are key to proper implementation of ESS:

- An optimum level of stress must be applied to the product to force defects into failure.
- The stress environment must not exceed the electrical or mechanical limits of the product, forcing needless failure or reducing the useful life of the product.

When these ideas are properly applied, ESS becomes a dynamic product reliability tool, with these attendant benefits:

- Reduced field repair expense
- Fewer defects and waste
- Elimination of less effective screening procedures
- Early detection of design and manufacturing flaws
- Improved production efficiency
- Lower unit costs
- Increased product value
- Improved customer satisfaction
- A better return on investment

The Evolution of ESS

Environmental stress screening has its origins in environmental testing methodology—basically, a means of exposing samples of product assemblies to one or more simulated field conditions. Driven by the military, environmental testing became prominent during World War II, when the proliferation of sophisticated weaponry, aircraft, and communications systems demanded a less cost- and time-intensive method of proving reliability than actual use testing. Laboratory experimentation with small-scale hardware was conducted. An insulated chamber, equipped with the technology to simulate environments such as temperature, humidity, altitude, and others, was used. These tests were performed during developmental stages to verify design and, on a manufacturing audit basis, to measure design compliance.

After the war, the new electronic technologies became available to the consumer market, a situation that created a different need. As products were downsized, their complexity increased, involving unique processing and assembly procedures. The accumulation of field failure data showed that design compliance was no longer sufficient evidence of reliability. It also led to the discovery that a number of unpredictable factors involved in parts design, product workmanship, and manufacturing processes were contributing to failure rates. As many of these were occurring during the infancy of the product, it was determined that testing methodology that could mimic the infancy stage of the product would provide the cure.

In an attempt to pass products through infancy, a process of powering products for an extended length of time was introduced. Referred to as burn-in, the process also generated a high degree of heat, a stress that many believed would have an added impact in precipitating early product failures.

Burn-in testing did succeed in forcing a small number of infancy failures, but not enough to significantly alter field failure rates. In addition, the process was time consuming, a factor that slowed production, increased unit cost, and delayed delivery.

An answer came, again from the military, with the introduction of its newly devised approach of environmental simulation testing in the form of military standards. These standards required that products be operated at specific environmental extremes. The logic was that field operation would demand that products be used in a variety of environments, as well as be exposed to cyclical changes within these environments. For example, a jet climbing from a desert runway to altitude would be exposed to a significant temperature change in a relatively short period. A static testing process, such as burn-in, could not provide the proper environmental simulation. To address this problem, a combination form of environmental simulation was introduced. This early application involved temperature cycling between hot and cold extremes paired with low-frequency vibration in a mechanical shaker.

These basic concepts led to the development of mission profile testing, in which products were exposed to environments that simulated actual use conditions. Mission profiling produced some important results. First, it was discovered that a number of products could operate successfully at the temperature extremes but would fail while being taken through multiple temperature cycles. This process, known as thermal cycling, was the basis for ESS theory. Second, a significantly higher number of failures was forced than with burn-in, particularly the types of failures that were occurring in the field. Finally, thermal cycling precipitated failures in a much shorter time. It also became evident that testing of this type produced a stress—through changing temperatures and the resultant expansion and contraction this caused—that would force previously undetectable latent defects into product failures.

From these discoveries, the concept of ESS was born, and the distinctions between this discipline and conventional test methodology became more defined. Where environmental testing was used primarily for design validation, environmental screening could be used to qualify product materials and workmanship.

As interest in ESS developed, a number of companies became involved in experimentation with different forms of stress and made attempts to establish specific guidelines for each type of program. As a result, some of the early distinctions drawn between ESS and conventional testing were blurred, creating confusion in industries where ESS had gained only preliminary interest. In an attempt to clarify the situation, the Institute of Environmental Sciences (IES) undertook a comprehensive study of ESS. Its book, *Environmental Stress Screening of Electronic Hardware (ESSEH),* set out to demystify the process and provide some general guidelines for ESS applications.

This effort, and follow-up studies conducted by the IES and other sources, provided solid evidence that ESS is a highly successful method of enhancing product reliability.

Misconceptions About ESS

Like any specialized technology, ESS is subject to misunderstanding. The following list includes the more common misconceptions:

- **ESS Is a Test**—ESS is not a test. It is not intended to validate design, which requires a failure-free simulation process. Rather, ESS is a screening process that requires stimulation to expose latent defects in products that would otherwise fail in the field.
- **ESS Is the Same as Burn-In or Aging**—ESS evolved from burn-in techniques, but it is a considerably advanced process. Burn-in is a generally lengthy process of powering a product at a specified constant temperature. ESS is an accelerated process of stressing a product in continuous cycles between predetermined environmental extremes.
- **ESS Involves Random Sampling**—ESS requires that all products be exposed. Because latent product defects are random by nature, screening all products is the only way to assure the effectiveness of the program.
- **ESS Can Be Used to Validate Design**—Although ESS may occasionally expose a design inconsistency, its intent and methodology are different. Design validation uses environmental simulation and laboratory analysis to determine whether a product is viable and whether it will work within the environment for which it was designed, usually called its mission profile. ESS, on the other hand, applies maximum permissible stimulation to products for the purpose of exposing latent defects related to parts, manufacture, assembly technique, or processing. ESS produces data that can be analyzed (e.g., for pattern part failures, process defects, or design errors), but its principal intent is to force failures. The mission profile, in this sense, is irrelevant. What is important is that the maximum degree of stress be applied to precipitate failure without detracting from the useful life of the product.
- **ESS Forces Needless Failures**—ESS is designed to force only those failures that would normally occur. ESS programs are based on a product profile that determines the maximum allowable stress the product can absorb without affecting its useful life. This determination is made prior to implementing the program by analyzing product function, tolerances, and failure rates. Ideally, there is never a loss of useful life. If there is, it will be negligible, particularly in light of the failures precipitated in-house.

- **All ESS Systems Are Alike**—ESS is as different as each product that is made. It is, by design, a product-specific program that must be tailored according to predetermined product variables. The type of screen, profile variations, fixturing, power supplies, inputs, outputs, and other considerations must be resolved before the program is implemented.

- **ESS Is Expensive**—ESS saves money and can deliver a fast payback. An ESS program is a cost that can and should be measured for its return-on-investment potential. By implementing the program properly, dramatic cost savings can be realized through reduced field warranty repair expense, lower unit costs, improved product value perception in the marketplace, and, ultimately, increased profitability. ESS is ideally implemented at a point in production where the cost to repair is lowest.

Types of Environmental Stress

Environmental stress screening consists of exposing a product to one or more stressful environments, such as thermal cycling, vibration, high temperature, electrical, thermal shock, and others. Profile characteristics, such as stress extremes, rates of change, amplitude, frequency, and duration must be tailored for each product. The goal in determining equipment requirements and profile characteristics is to achieve maximum screening strength within the realm of product limitations and overall affordability. Useful screening strength information is available in the U.S. Department of Defense document DODHDBK-344 (USAF). A description of some of the more commonly used environments follows.

Temperature Cycling

Temperature cycling consists of multiple cycles of changing temperature between predetermined extremes. Figure 7.3 shows a typical temperature profile.

Because all variables in a screen are product dependent, temperature extremes must stop short of damaging the product but must be far enough apart to allow for optimum stressful lengths of temperature change between the extremes. Cycle times must follow this same rule of thumb, because the constant rate of change provides the expansion and contraction necessary to sufficiently stress the product.

The product temperature change rate is dependent on specific heat properties of the product, the difference between product and air temperatures, and surface conduction factors involving air velocity and direction. Assuming proper air flow, the temperature of a typical circuit board would resemble Figure 7.4 when subjected to a typical temperature profile.

Because the dwell period at the temperature extremes does not significantly contribute to the stress, many manufacturers allow the product to remain at temperature extremes only long enough to allow a functional test of the product, as shown in Figure 7.5.

Another step some manufacturers have taken to maximize the stress during thermal cycling is to adjust the chamber air temperature so that the high and the low temperatures are close to the extremes the product can withstand. However, in researching this stress profile, it must be

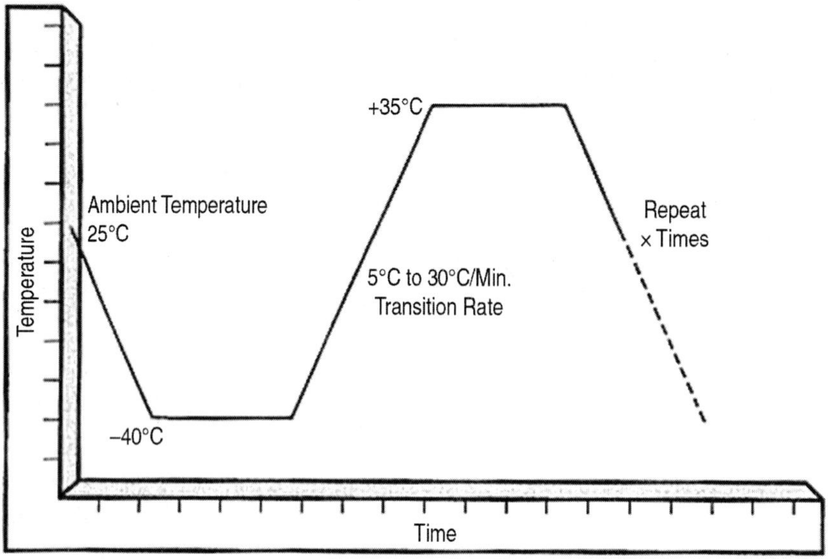

Figure 7.3 *Typical ESS temperature profile.*

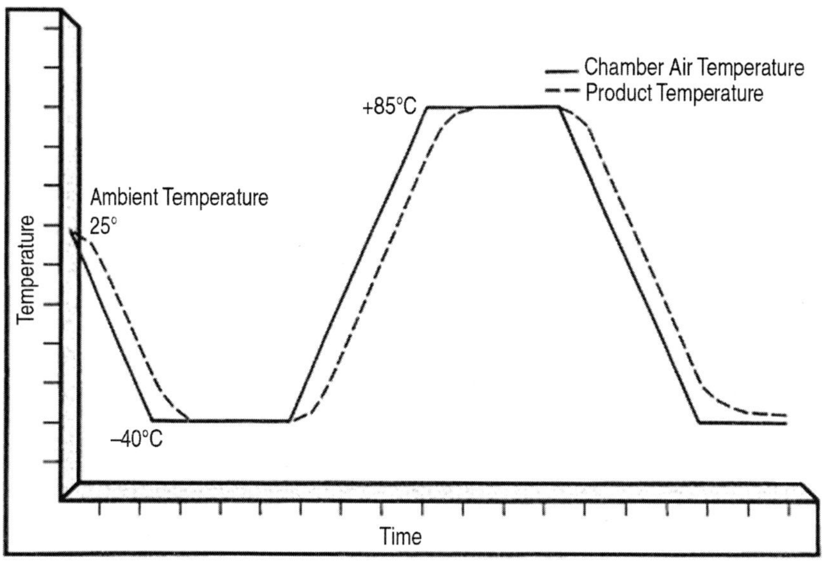

Figure 7.4 *Chamber and product temperature profile.*

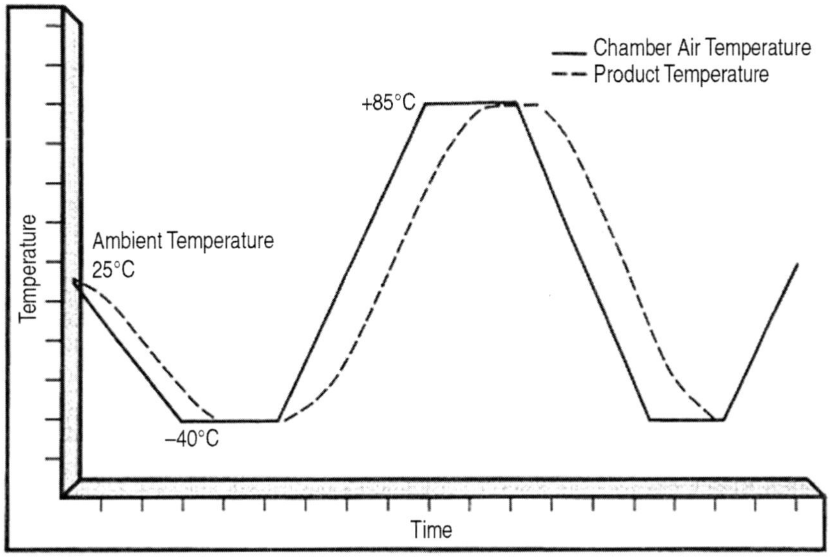

Figure 7.5 *Temperature profile with functional test at temperature extreme.*

determined that multiple cycles between these two extremes do not erode the useful life of the product. Figure 7.6 illustrates such a profile.

Air flow in a thermal cycling chamber is critical because it directly affects product temperature change rates. Figure 7.7 illustrates product temperature during four separate runs of a profile

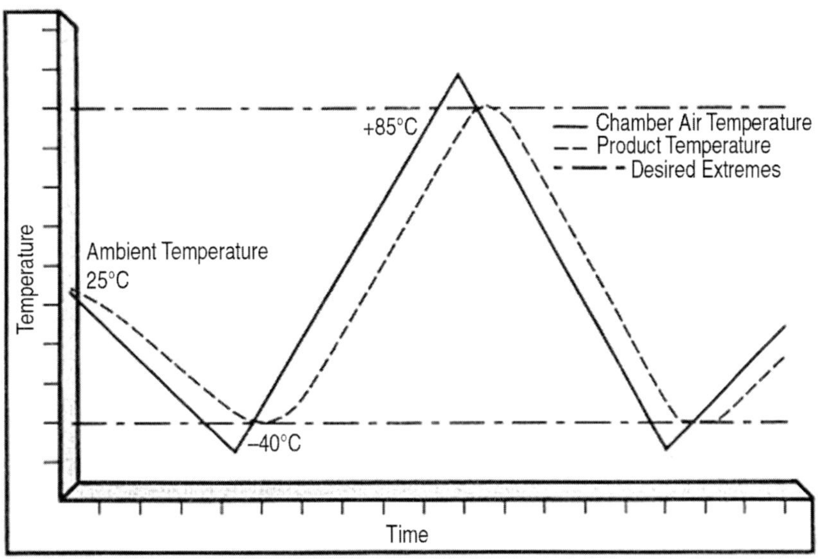

Figure 7.6 *Temperature profile near product destruct limits.*

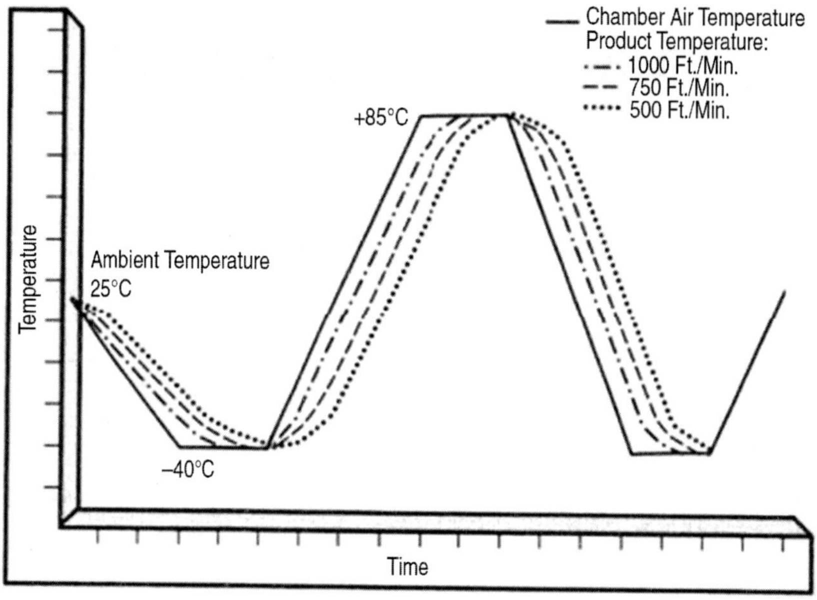

Figure 7.7 *Temperature profile with air velocity.*

that had identical air temperature change rates. The only difference between the runs is the air velocity. Clearly, the chart shows that the lower the air velocity, the slower the product temperature increases and decreases, and less stress is brought to bear on the product. The higher air velocities cause a more rapid product temperature change; thus, these higher air velocities subject the product to a higher degree of stress.

The optimum air flow depends on the product. There is an air velocity at which maximum heat transfer is obtained. Exceeding that air velocity can be counterproductive. The air velocities listed in Figure 7.7 may or may not be the most appropriate for the product. The correct air velocity rates and air direction for a particular product can be determined through experimentation.

The general consensus is that thermal cycling is the most effective screen. Similar to any screen, it must be properly implemented. Failure rates must be analyzed to determine which latent defects are causing failures, and experimentation must be performed to determine the screen profile best suited to trigger those particular latent defects into failures.

Random Vibration

Considered the most effective of three principal types of vibration, random vibration involves the excitation of a product with a predetermined profile over a wide frequency range, usually from 20 to 2000 Hz. Product stress is created through simultaneous excitation of all resonant frequencies within the profile range.

Random vibration is applied by mounting or attaching a product to a unit called an electrodynamic shaker, controlled by a closed-loop digital system and dedicated fixturing. Fixturing must be extremely rigid to ensure that stress is transmitted directly to the product and that the process is repeatable with reasonable accuracy. Products may be vibrated on a single axis or on multiple axes concurrently or consecutively. Currently, opinions vary as to which provides the more stressful environment.

Random vibration screens generally require less time to run than other ESS programs and are considered particularly effective in exposing mechanical defects, such as loose solder, improper bonds, and printed circuit board (PCB) shorts.

The primary drawbacks are equipment cost and lack of screen uniformity. Regarding the former, electrodynamic shakers may be expensive to install, control, and maintain. Random vibration also is, by nature, less effective than temperature cycling in providing a uniform stress environment. To avoid overstress, random vibration screens are tailored to exert maximum stress on parts at joints located at the center of a product and are gradually reduced for parts closer to the edge, a factor that may result in failure to expose all potential defects.

Nonetheless, random vibration is considered superior to other forms of vibration and is often used, where financial objectives allow, in conjunction with thermal cycling screens.

High-Temperature Burn-In

This process is generally a static one in which a product is subjected to elevated temperatures over a predetermined period. This screen evolved from the idea that continuous operation of a product would force infancy failures. It also was believed that providing additional heat would escalate the rate of failure. It is more likely, however, that increases in screen effectiveness are related to changing temperatures during heat-up and final cool-down, rather than powering at a constant high temperature.

Electrical Stress

This process is used to exercise circuitry and simulate junction temperatures on semiconductors. There are two basic types:

1. Power cycling, which consists of turning product power on and off at specified intervals

2. Voltage margining, which involves varying input power above and below nominal product power requirements

Electrical stress does not, according to research, expose the number of defects commonly found through thermal or random vibration screens; consequently, it is considered much less effective. It can, however, be relatively inexpensive to implement electrical stress with another screen to increase overall screen effectiveness. It also may be necessary to power products in order to find soft failures.

Thermal Shock

Thermal shock is a process that exposes products to severe temperature extremes, usually in rapid fashion. That is, a product is continually transferred—either mechanically or manually—from an extremely hot environment to an extremely cold environment and back. Thermal shock is generally considered a cost-effective way to screen for defects at the component level, particularly integrated circuit (IC) devices that require a high degree of stress to experience the rates of change needed to force latent defects into failure.

Thermal shock also may be useful at other levels of assembly, as long as the severity of its rates of change do not cause needless damage. This is a particular risk with more complex assemblies, such as those containing components other than ICs. Of equal consideration is the cost efficiency of a thermal shock screen. Generally, the equipment needed to provide an effective screen is expensive. In addition, because products must be shifted from one environment to another, there will always be an unused chamber. If manual transfer is involved, the risk of accidental product damage increases. Finally, a thermal shock screen is difficult or, in some cases, impractical to power or monitor. This limits opportunities for accumulating field failure analysis data.

Sine Vibration, Fixed Frequency

This is a form of vibration that operates on a fixed-sine or single operational frequency. This method usually requires a mechanical shaker with a frequency range up to 60 Hz. Although less expensive and easier to control than a random vibration screen, fixed frequency vibration is generally viewed as not providing as effective a level of stress.

Low-Temperature Screen

Similar in concept to a high-temperature screen, a low-temperature screen is based on the principle that failures will be forced by the contrast between the heat generated by powering a product and the cold environment.

Sine Vibration, Swept Frequency

This is a form of vibration that operates on a swept-sine or multiple operational frequency. It usually requires a hydraulic shaker with frequency range up to 500 Hz and is viewed similarly to fixed frequency in terms of overall effectiveness.

Combined Environment

Depending on product complexity, cost, and reliability specifications, environmental screens may be used in concert with each other. For example, thermal cycling and random vibration are often combined in an ESS program. The primary considerations should be whether an additional stress—applied simultaneously or consecutively—will expose a significant number of additional defects, and whether the cost of the additional stress is justifiable.

In general, the types of stress used and how the stress is applied are dependent entirely on what will precipitate the greatest number of failures in a specific product in the shortest period of time. As with all screens, the most effective profile is product dependent. However, some recommendations are available in the IES *Environmental Stress Screening of Electronic Hardware (ESSEH)* guidelines, the Rome Air Development Center (RADC) studies, and the U.S. Navy Manufacturing Screening Program, NAVMAT P-9492.

Advantages of Temperature Cycling

A number of independent studies have been conducted that analyze the effectiveness of various types of ESS processes. Organizations such as the IES and the RADC have compiled in-use data that conclude thermal cycling is the most effective type of screen, as shown in Figure 7.8. A comparison in weighted rank of effectiveness among various environments is shown in this table published by the IES in its *Environmental Stress Screening of Electronic Hardware (ESSEH)* guidelines.

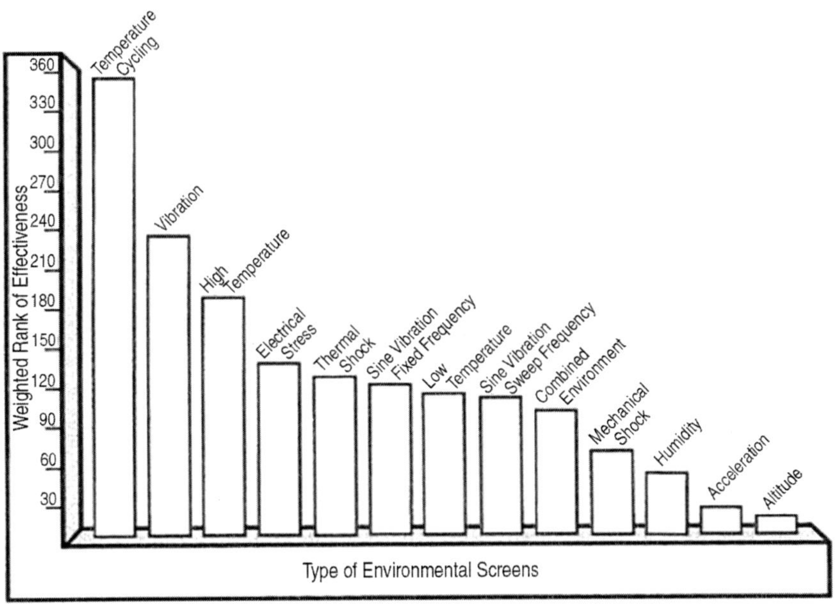

Figure 7.8 *ESS screen types.*

Although it has been shown that the most effective screening program may involve more than one screen, thermal cycling is considered the single most effective screen in terms of identifying latent defects over a period of time. According to the IES *Environmental Stress Screening of Electronic Hardware (ESSEH)* guidelines published in 1981, 1984, and 1988, thermal cycling, when compared with random vibration (ranked second most effective), regularly detected an average of two-thirds more latent product defects.

Thermal cycling provides the additional advantage of a more uniform stress environment—one in which all areas of the product are subjected to a relatively more equal amount of stress throughout the screen profile. Some other environmental stresses cannot, by their nature, provide uniformity and have a higher inherent risk factor.

Stress uniformity also makes thermal cycling easier to control, permitting greater flexibility in implementation, revision, or refinement.

Finally, thermal cycling takes less time to perform than most other types of stress. (It is less time consuming than any form in terms of its ability to force the greatest number of defects into failure. Random vibration, which has a shorter cycle time, does not provide the same overall efficiency.) The net result is better product throughput with minimal impact on production.

The basic concept of thermal cycling involves three uncomplicated principles. First, the temperature extremes must be as far apart as possible—regardless of the intended operational limits of the product—as long as they do not cause needless damage. Second, the rate of change between the extremes must be as rapid as possible—usually greater than or equal to 5°C per minute—to create the optimum level of stress, again without damaging the product. Figure 7.9 shows a comparison of various temperature change rates and their impact on screen effectiveness.

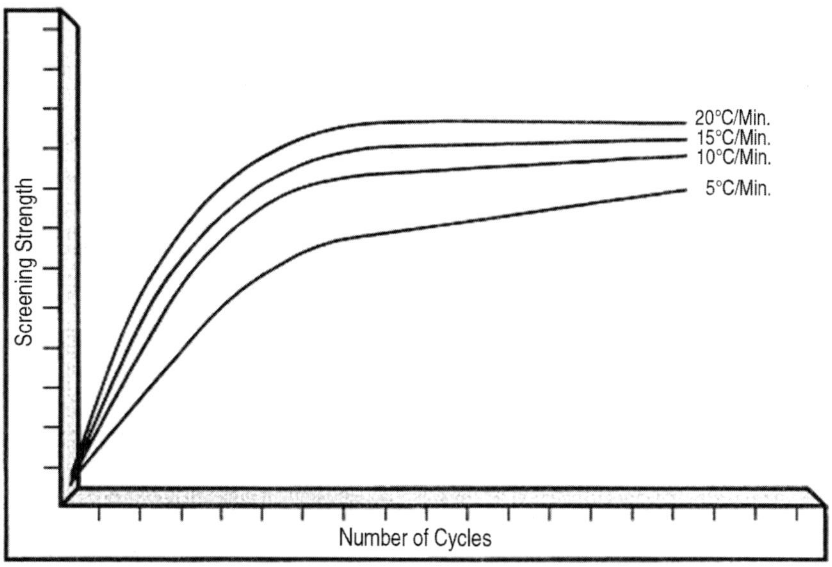

Figure 7.9 *Effect of temperature on screen effectiveness.*

Finally, the temperature of the product must follow, as closely as possible, the temperature inside the chamber. This is achieved by assuring that high-velocity air is properly routed over the product. Figure 7.7 illustrates how rapidly the product temperature changes at various air velocities. Clearly, the temperature of the product changes faster as the air velocity across it is increased. Faster temperature change creates greater stress on the product; consequently, it can more quickly force latent defects into failure.

A good starting point to determine an effective thermal cycling stress profile would be to start with the following:

- Temperature extremes equal to or greater than a 100°C temperature difference between hot and cold
- Product temperature rate of change equal to or greater than 5°C per minute
- Air velocity of approximately 750 fpm (feet per minute) at the product
- Enough cycles to effectively cause latent defects to produce failures

Levels of Screen Complexity

Making ESS work to your best advantage requires consideration of whether it is either necessary or justifiable to augment the primary stress environment. Certain products may or may not benefit from electrical stress, such as powering or signal generation, during the screen. Others may require it.

To make this determination, some basic considerations are necessary. The environment in which the product must operate is important. A device that will be used in a controlled environment, such as a computer, will most likely not require special treatment. Shipboard instrumentation, on the other hand, may be exposed to a volatile range of environments and most likely would require more than a basic approach.

Beyond environment, different screening levels can provide information relating to types of failures. It may be necessary to know whether a failure is hard (i.e., a point at which the product will stop functioning) or soft (i.e., a situation in which a product will cease functioning under certain conditions but will operate under others). Hard failures are always defined as critical, because the product is unable to operate under the conditions in which it is expected to operate. Soft failures can be critical if they occur within intended operational specifications of a product. Amplified levels also can provide failure analysis data that indicate when and where a failure occurred. Finally, various levels of power cycling create additional stress that can be beneficial in forcing latent defects to cause failure.

The overriding consideration is to provide as much stress as is cost justifiable to gain the desired result. Once this need has been established, the appropriate level of screen complexity can be factored into the program. The basic types are described as follows:

- **Static**—This means subjecting an unpowered product to a single stressful environment. The product is typically tested before and after the screen to verify operation.
- **Dynamic**—Power is applied to the product during the screen. Another type of stress may be added to the product at the same time the power is being cycled on and off. Also,

power levels can vary above and below nominal values. Power is usually turned off during hot-to-cold transitions so that heat generated by the product does not affect descending (i.e., pulldown) product and chamber temperature rates, which reduces stress during that part of the screen.

- **Exercised**—Power and the additional stresses of supplying inputs and monitoring or loading outputs are applied to the product. In some cases, signals or loads must be applied to a powered product to avoid damage.
- **Full Functional**—The product is operated as though it were in actual use. This can include power and all necessary signals and loads, which may or may not be cycled on and off.
- **Monitored**—This is a description of any powered product screen. It consists of monitoring the product for failures and is used in indicating, for example, products that fail and then resume operation (i.e., soft failures), the conditions at the time of failure, and the time of failure. This type of information is critical in cases where even soft failures cannot be tolerated. It also can be used to identify and correct product or process deficiencies, or to fine tune the screen profile.

Failure Analysis

Once an ESS program is implemented, its long-term success depends on monitoring the results. This, in fact, can often be the single most critical factor in determining whether ESS delivers the proper return on investment (ROI). New suppliers, a change in process technology, new assembly technology, or a sudden drop in in-house failure rates without a corresponding drop in field failure rates are all reasons for reviewing a program.

Data acquisition is essential to the process because it helps establish the learning curve required to keep personnel integrated with program objectives by providing the following:

- Failure symptoms
- Environment at the time of failure
- Cycle during which the failure occurred
- Time into the cycle during which the failure occurred
- The cause of failure, based on analysis of the failed part or assembly

Analysis of this data should be grouped into major failure categories and should be plotted against the screening cycle and the time at which the defect was found. As data accumulate, problems can be more readily identified and resolved, and screen profiles can be altered accordingly.

Observing sound failure analysis practice goes beyond establishing the initial ESS profile; in effect, it creates a closed-loop system that builds accountability into the program and allows for change based on hard data, rather than supposition.

Case Histories

The IES, RADC, and other groups have researched ESS in detail, compiling studies and case histories relating to methodology, combination screening, screen complexity, cost analyses, and other factors.

In addition, a number of companies have independently written about their success with ESS programs. A few examples are detailed in the following paragraphs.

Analogic Corporation

Analogic Corporation found an excessive number of early-life or infancy failures in one of its products, even after steady burn-in screening. Detecting no inherent design problems, company executives applied an ESS program of thermal cycling to the product under power in an attempt to detect a greater number of defective products.

The thermal cycling reduced the final-assembly-level reject rate to less than half the rate achieved on the same line in the preceding year. This reject rate included a significant increase in the number of defects identified in integrated circuits and displays—defects that did not appear in the previous steady-state burn-in screening. In addition, thermal cycling shortened the necessary aging time and reduced production time. As a result of these findings in the laboratory, Analogic installed six thermal cycling systems on its production lines.

Bendix Corporation

Bendix Corporation, makers of electronic fuel injector systems from 1975 to 1980, published some significant findings about its experience with ESS technology. In designing production lines for its injector systems, Bendix wanted to ensure product reliability because an injector system field failure could disable a car and result in expensive field repairs. Because the product was extremely complex (involving 420 components and 1,700 soldered joints), the conventional screening methodology was considered inadequate. Bendix also wanted screening to be more effective in identifying defects that caused intermittent, or soft, failures.

Bendix replaced the traditional burn-in screen on the production line with thermal cycling. The ESS system proved more effective in spotting the solder, workmanship, and semiconductor defects that contributed to soft failures. As a result of using thermal cycling, field failures caused by these defects were reduced from 23.5 to 8%.

Hewlett-Packard (HP)

Hewlett-Packard (HP) introduced its 9826A desktop computer line in 1983. To be competitive, the company needed to spend less on production and still produce an extremely reliable product.

The company installed an ESS program consisting of a thermal cycling screen. The screen enabled the company to cut back production-line screening time from the two to five days needed for a typical burn-in screen to a few hours. The thermal cycling screen prevented a production-line bottleneck, increased productivity, and reduced the unit production cost.

Because the ESS screen was applied very early in the production phase on relatively simple subassemblies, the company also spent less time on troubleshooting and repairs. There was also a substantial reduction in the field warranty rate. After six months, thermal cycling was exposing two to three times more latent defects than similar production lines. According to a six-month study, ESS directly contributed to a 50% drop in the number of field repairs made under warranty. Research shows that total savings to the company will amount to more than $1.5 million over five years, after the cost of installing the ESS program.

In addition, ESS has become a powerful tool in the automotive industry, where "the Big Three" automakers have embraced the concept as readily as statistical process control (SPC). In aerospace, companies such as Motorola, Inc., Hughes, Inc., and Boeing have successfully implemented programs. Many consumer electronics companies also have used ESS to reduce unit costs and to gain a competitive edge.

Implementing an ESS Program

To be successful, ESS must be treated as a highly technical discipline. It must involve careful research, a sound set of objectives, and a system for monitoring ongoing effectiveness. The basic steps are described as follows:

1. **Establish an Objective**—The reasons for undertaking any form of ESS must be clear for the program to work. Some principal considerations are as follows:

    - Achieve and maintain high field reliability by finding latent defects in-house.
    - Reduce production costs by exposing and correcting defects early in the manufacturing process.
    - Meet a contractual reliability demonstration requirement.
    - Assure cost effectiveness in a reliability improvement warranty (RIW) contract.
    - Reduce field costs of operations and maintenance.
    - Upgrade in-house reliability methodology.

2. **Involve Key Personnel**—Essential to program planning is that technical personnel from all applicable departments be involved. This in-house team may consist of, but not be limited to, representatives from engineering, production, testing, and quality. For example, quality can examine field failure histories and report the type of latent defects that cause field failures. These are the defects that must also precipitate failures in-house for ESS to be effective. Engineering can provide information on product tolerances, handling limitations,

and other factors. Production can point out areas of process complexity. Testing and field services can assist in selecting the most appropriate screen profiles to expose latent defects, as well as develop field failure expectancy rates for new products.

3. **Involve a Supplier**—Selecting the right supplier can make an important difference in both the cost and effectiveness of an ESS program. This may include being able to provide assistance in clarifying program goals, as well as the equipment information necessary to develop cost analyses. The latter point is of particular importance because the need for, and the cost of, future system integration will be determined by how complete a program the supplier can initially deliver. A credible supplier should also be able to provide appropriate agency approvals for equipment and be able to support equipment, both in terms of warranty and service.

4. **Analyze the Business Impact**—Based on information from all applicable departments, management can then determine whether an ESS program is justifiable. This can and should be evaluated on a return on investment (ROI) basis, involving considerations such as the following:

   - Will the reduced field maintenance cost justify the program cost?
   - Is ESS necessary to eliminate excessive latent defects?
   - Is ESS necessary to meet a reliability requirement?
   - Will ESS save money in production?
   - Is an improved production schedule worth the cost of the program?
   - Will the good will and increased sales potential from delivering latent-defect-free products balance the cost of the program?

5. **Determine the Best Level for the Program**—Implicit in the cost advantages of ESS is the capability to force failures at the most cost-effective point in the production process prior to shipment. Figure 7.10 illustrates the basic production levels, including cost-to-repair values taken from a typical industry example and a level of complexity ratio that is fairly common throughout the industry.

   Although logic might suggest that screening at the relatively inexpensive component level is best, also note that screening at such a low level of complexity may not produce meaningful results. Screened components, for example, can still be damaged during assembly. On the other hand, looking at the board level, there is a level of complexity very close to final assembly, but at a significantly lower cost to repair. Depending on the levels of complexity involved at each stage of production, it may be justifiable to screen at more than one level.

   In the case of the company cited in this example, the decision to screen at the board level was made. The early identification of defects greatly reduced field repair costs, a savings that quickly paid for the ESS program and later translated into increased profitability.

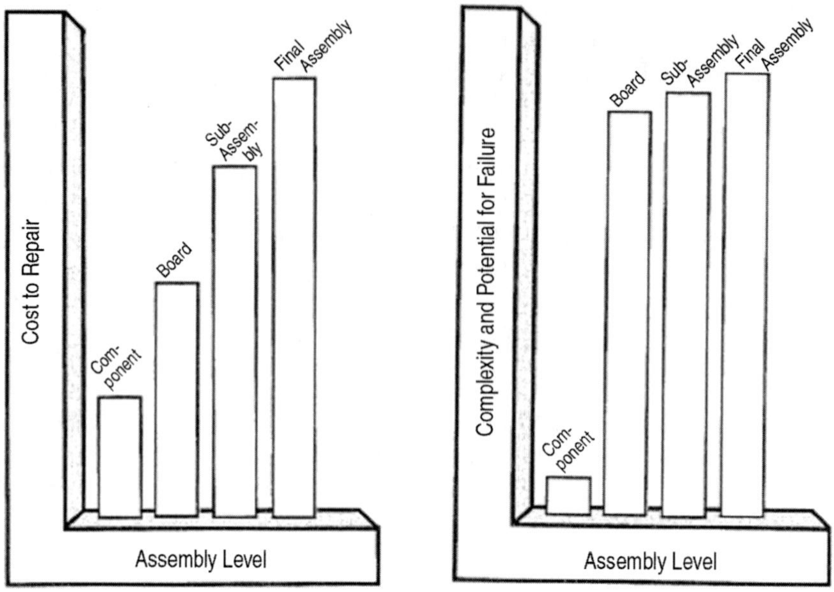

Figure 7.10 *ESS failure versus production level.*

In each case, it is important to realize that a certain number of trade-offs are implicit at each level of screening and that careful analysis of these variables will help determine where a program will be most effective.

- **Tailor ESS to the Product Requirements**—ESS is a product-specific program, requiring a screen or set of screens that will provide the highest level of stress without diminishing product life. To achieve this, each product must be treated differently in establishing a screening profile. Its operating requirements, in-use environment, and failure history must be used to develop the proper program.
- **Establish a Method for Monitoring the Program**—Follow-up in ESS is not only good practice—it is essential. Because each program is based on field failure analyses, it is only logical that this would continue throughout the program. This assures the program is performing as specified. In addition, monitoring provides a basis for assessing change—from new parts and processes to a falloff in screen effectiveness—and responding with appropriate action.

Equipment

Because an ESS program must, by nature, be tailored to meet a specific set of objectives, it is important that equipment be considered in relation to the primary intent of ESS: to provide optimum stress in the shortest amount of time without damaging the useful life of the product.

Keeping the product always as the focal point, it is possible to take a logical approach to equipment specification, as illustrated in Figure 7.11.

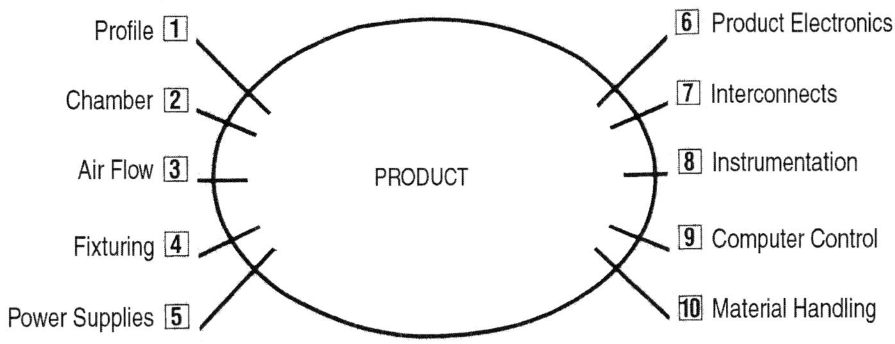

Figure 7.11 *ESS program equipment defined by the product.*

1. **Profile**—This will determine which types of equipment and accessories are required, based on an examination of screen types, screen complexity, and research and analysis of field failures.

2. **Chamber**—The major considerations associated with selecting a chamber are as follows:

    - The type and size of the product to be screened
    - The quantities to be screened at any one time
    - The number of interconnects required to and from the product, if it is to be powered
    - The temperature range and change rates of the screens to be performed
    - The data transfer rates of the product, if applicable
    - Air flow, velocity, and direction to maximize product heat transfer
    - Product handling that can affect throughput and product reliability

    Chamber construction is usually contoured to the types of stress and amount of heat that will be generated. This may or may not include complex fixturing, signal lines, and different types of carriers to hold or load product. Of consideration is that a minimum of product handling should be involved because this can increase the risk of product damage outside screening and minimize the effectiveness of the program.

3. **Air Flow**—To ensure proper screen effectiveness, equipment should be designed to provide adequate air velocity and uniform air distribution. A function of both chamber design and equipment placement, the objective is to assure that the same temperature change rates are occurring on all products so that the full impact of the thermal cycle is realized.

4. **Fixturing**—Fixtures are necessary to hold products in the proper orientation to benefit from the screening process. Fixtures may also be designed to allow for easier handling and for compatibility with power supplies and interface electronics in situations where different levels of screen complexity are involved.

5. **Power Supplies**—Power supplies are necessary to apply power cycling.

6. **Product Electronics**—Electronics are used to create an interface with a product when it must be exercised or monitored and may include input and output (I/O) lines, sensor devices, and instrumentation connections.

7. **Interconnects**—These bring signal and power lines outside the chamber and must be compatible with available hookups and power sources.

8. **Instrumentation**—Various types of instrumentation are available for system and product safety and operation. Consideration must be based on need, as well as the potential for upgrading the data acquisition capability of a system.

9. **Computer Control**—Computer control can be used to control total program operation and monitoring. Off-the-shelf software programs are available, but it is still a buyer-beware situation. Many programs are too general for specific applications or are too complicated for personnel to use. A customized, user-friendly package should be sought.

10. **Material Handling**—This is an occasionally overlooked but decidedly important aspect of a program, particularly with large products involving a high level of screen complexity. Where cost effective, devices that automatically load or unload product or that facilitate manual handling of product should be used. This reduces the possibility of causing needless product damage outside the screening process.

Burn-In Optimization

The use of environmental stress tests for burn-in can be expensive. In some cases, more factory floor space is dedicated to burn-in than for assembly. Burn-in costs can be minimized by optimizing the burn-in duration. There are three components to burn-in costs:

1. Laboratory costs
2. Laboratory failure costs
3. Field failure costs

Burn-in is effective only if the failure rate is decreasing. Referring to Figure 7.12, if a product is burned-in for the period from time = 0 to time = x, the product will be impacted in one of three possible manners:

1. If the failure rate is increasing, the failure rate at time = x is greater than the failure rate at time = 0. Burn-in has reduced the reliability of the product.
2. If the failure rate is constant, the failure rate at time = x is equal to the failure rate at time = 0. Burn-in has had no effect on the reliability of the product.
3. If the failure rate is decreasing, the failure rate at time = x is less than the failure rate at time = 0. Burn-in has improved the reliability of the product.

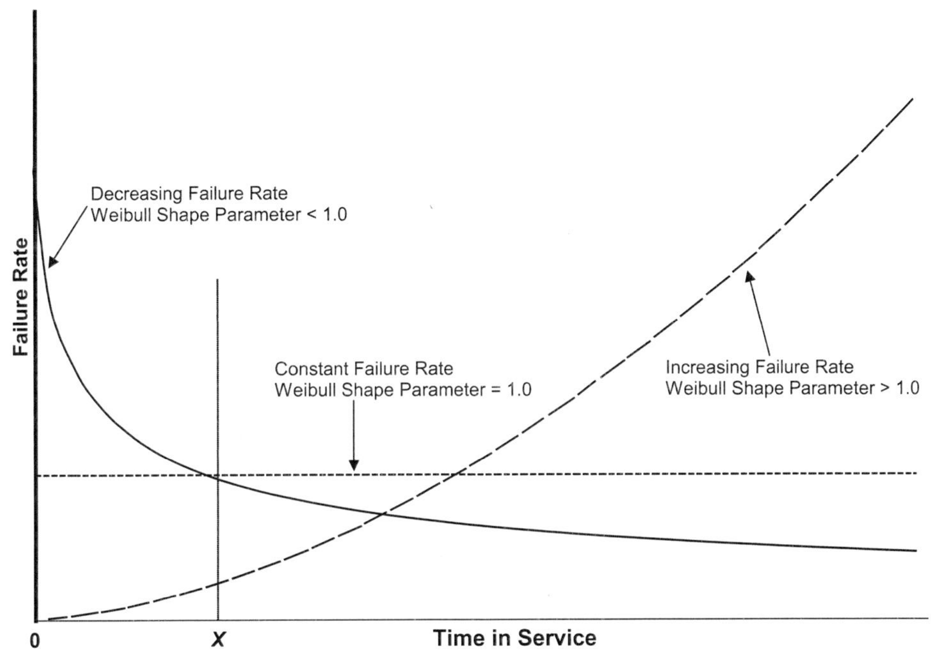

Figure 7.12 *Failure rate characteristics.*

The total cost to burn-in a component or assembly is

$$C = Ax + L\left[1 - R(x)\right] + F\left[1 - \frac{R(x+w)}{R(x)}\right] \tag{7.1}$$

where

A is the cost of burn-in per unit time

L is the cost of a failure during burn-in

F is the cost of a failure in the field

w is the warranty period (beyond this time, there is no field failure cost)

Using the Weibull reliability function, the total cost to burn-in a component or assembly is

$$C = Ax + L\left[1 - e^{-\left(\frac{x}{\theta}\right)^{\beta}}\right] + F\left[1 - \frac{e^{-\left(\frac{x+w}{\theta}\right)^{\beta}}}{e^{-\left(\frac{x}{\theta}\right)^{\beta}}}\right] \tag{7.2}$$

Example 7.1: An electronic component has a Weibull time-to-fail distribution with a shape parameter of 0.6 and a scale parameter of 140,000. The warranty period for this component is 8,760 operating hours. A field failure costs \$900, and a failure during burn-in costs \$25 to repair. Burn-in costs \$0.037 per hour per unit. What is the optimum burn-in period?

Solution: The optimum burn-in duration is found by minimizing the following equation in terms of x:

$$C = 0.037x + 25\left[1 - e^{-\left(\frac{x}{140{,}000}\right)^{0.6}}\right] + 900\left[1 - \frac{e^{-\left(\frac{8{,}760}{140{,}000}\right)^{0.6}}}{e^{-\left(\frac{x}{140{,}000}\right)^{0.6}}}\right]$$

From Figure 7.13, it can be seen that the optimum burn-in duration is 163 hours.

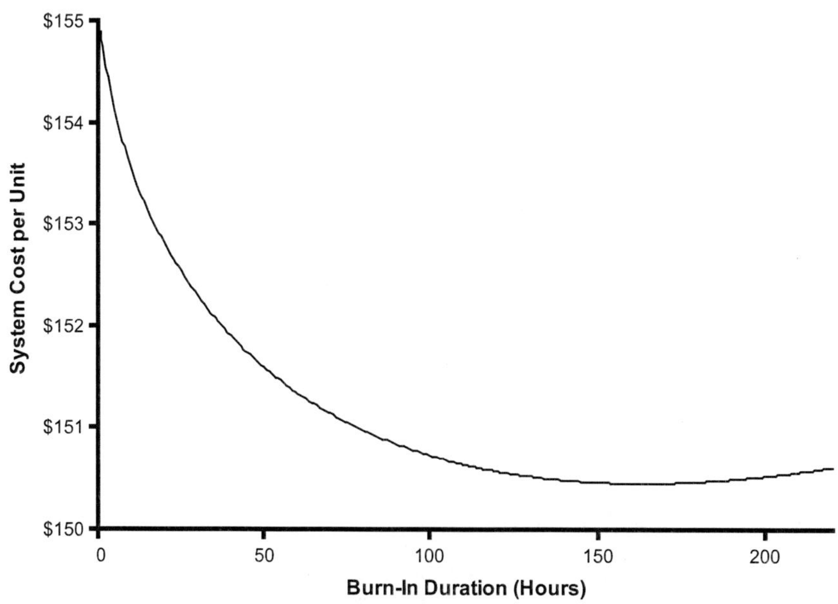

Figure 7.13 *Graphical solution for burn-in duration.*

Summary

ESS is an important consideration for any manufacturer with an unacceptable field failure rate caused by production process problems, random defective components, or other problems not related to product design.

ESS is a comprehensive program for improving product field reliability because it involves 100% of product and is structured to force failures of latent defects in-house, instead of in the field. It will achieve this end at a much more effective and faster rate than conventional aging methods.

The benefits of ESS include increased profitability by lowering the number of warranty repairs, reduced costs, and improved product and corporate market value perception.

Before implementing an ESS program, a careful analysis involving key technical personnel is necessary to ensure that the program is both performance and cost effective. Continued in-house failure analysis is vital to maintaining program effectiveness.

An ESS program should be tailored to meet specific product requirements and to subject the product to stresses that will force the maximum number of failures without altering the life expectancy of the product. Equipment specification and implementation should follow this logic and be installed by a supplier able to explain, justify, and support the equipment.

Finally, as with any investment, ESS must pay for itself. A properly considered and applied program will generate a measurable ROI—through improved product quality, reduced failure rate, and increased profitability.

Chapter 8

Test Equipment Methods and Applications

The basis of all environmental testing is to evaluate a product based on its mechanical failures. Either the product survives a defined test without failures, or failures precipitated by the test are evaluated for product improvement. The field of environmental testing is very broad, covering many different disciplines. Vibration and thermal cycling are the two most commonly used test disciplines; however, the operating requirements of a product frequently necessitate the use of other environments such as humidity, salt fog, sand and dust, and electromagnetic interference (EMI). Often, the best test method is to simultaneously combine the applicable environments.

The testing process is begun by determining the environments to which the product will be subjected and then evaluating which of these environments are critical to the survivability of the product in its real-world usage. Next, data are acquired from these environments so that the environments can be simulated in the laboratory. This is then followed by development of the laboratory test and its application to the product. Every detail of each and every step of this process is critical for success. Any incorrect assumptions about the consistency of the real world, any improperly acquired or reduced data, or a poorly applied test will culminate in less-than-reliable test results.

Obviously, the best test for any product is to subject it to the real-world environment for its lifetime. This is usually impractical and excessively expensive. For example, an electronic box that is mounted in an automobile would best be tested by driving the automobile for many thousands of miles over different road conditions in cold and hot climates and all weather conditions. The resultant test would take months and destroy an entire automobile in the process. The goal of laboratory testing is to achieve results that are identical to those of the real world but at a lower cost and often with a reduced time.

The first step is to define the real world for the product. Where, when, and how will the product be used? If it is an electronic entertainment center designed for use in the home, its operating environments will be considerably less severe than those of the electronic box in an automobile. The entire life cycle of the product must be considered—from the time its construction is completed until its desired operational lifetime ends. This includes shelf storage and shipping before final usage. It is possible that shocks and thermal cycles that can be induced by handling and long-term storage of the product in its shipping container are critical to the subsequent reliability of the product.

Serious consideration must be paid to determining what is the real world for the product. Temperature extremes, numbers of thermal cycles, humidity levels, and salt fog conditions can be evaluated from the meteorological conditions in the geographic regions where the product will be located. Electrical environments are based on the operating conditions. Shock loads can be developed based on handling and operation. However, vibration is a much more difficult environment to determine, evaluate, and test. The size of the product, its orientation to the input levels, the applied frequency range, the critical frequencies of the product, and the available testing equipment all must be considered in the development of the laboratory vibration test.

With a product located in an automobile, there is a great range of environments. The vehicle must be able to operate in both arctic and tropical conditions, at altitudes from sea level to those in high mountain ranges, in humidity levels from the desert to the salt fog at the seashore, and over an infinite variety of road surface and speed conditions. All of these must be considered, evaluated, and included in the laboratory testing environments.

Three classifications of testing are commonly used:

1. Defect exposure
2. Acceptance
3. Qualification

The purpose of defect exposure tests is to evaluate the product during its assembly to ensure that it has been manufactured properly. The most commonly used type of defect exposure test is environmental stress screening (ESS) (see Chapter 7, Environmental Stress Screening), which uses vibration and/or thermal cycling, with the test environments being based on standard procedures rather than on the intended environments of the product or on correlation with the real world. Acceptance tests are used to evaluate production hardware and ensure that it meets the desired requirements and is ready for installation and use in the final product. Qualification tests are used for evaluating development hardware and initial production hardware, and usually simulate the entire lifetime of the product.

All of these tests are critical to developing and producing cost-effective, reliable products. Considerable effort must be placed on developing and performing these tests so that they are accurate and effective.

Simulation Versus Stimulation

Simulation is an attempt to reproduce usage conditions, whereas stimulation testing is conducted in the absence of real-world data or to rapidly precipitate failures.

Simulation

Simulation is an attempt to match the real-world environment in the laboratory. The real-world environment must be defined and then data acquired while the product is being subjected to it. By nature, the laboratory environment approximates the real world as closely as possible, within the limitations of data acquisition, data reduction, laboratory equipment capabilities, and practical limitations (e.g., schedule).

Application

Simulation testing is applicable to every product that must withstand cycling environments during its usage. This includes thermal, humidity, and electrical cycling, as well as shock and vibration and other environmental conditions.

Typical Test Types

Because of the broad spectrum of physical environments, there are many typical types of simulation tests. For large items such as full vehicles (e.g., cars and trucks), a laboratory vibration test would consist of multiple axes inputs at the wheel locations, using real-time methods. The electronics in the vehicle can better be tested using single-axis electrodynamic vibration, where their critical frequencies can be applied.

The optimum laboratory test matches the real world as closely as possible. Because the real world includes the simultaneous application of multiple environments, these should be included in the laboratory test whenever possible. To this end, environmental chambers can be used over shakers to simultaneously apply thermal cycling and humidity along with vibration and shock.

Defining the Real World

The first critical step in developing a laboratory simulation environment is to define the real world that is being simulated. By necessity, the following can give only general guidelines for defining the real world, because the environments experienced by hardware vary greatly. These guidelines give an indication of what must be considered. Defining the real world is a difficult process and can require considerable research. It includes everything that happens to the hardware after it has been manufactured. Table 8.1 can be used as a guide in considering the requirements for defining the real world. This table lists the possibility of a specific environment occurring during a phase of the life of the product. The three phases are as follows:

1. Storage
2. Shipping and handling
3. Customer usage

The following are the three categories of applicability:

1. Rarely
2. Occasionally
3. Usually

These are relative terms to assist with determining a real-world definition. They refer to how commonly the environments are experienced by a general spectrum of products. "Usually"

TABLE 8.1
REAL-WORLD ENVIRONMENTS

Environment	Product Life Phase		
	Storage	Shipping and Handling	Customer Use
Thermal:			
High Temperature	Occasionally	Usually	Usually
Low Temperature	Occasionally	Usually	Usually
Cycling	Rarely	Usually	Usually
Shock	Rarely	Occasionally	Occasionally
Mechanical:			
Shock	Rarely	Usually	Usually
Vibration	Rarely	Usually	Usually
Electrical:			
Voltage	Rarely	Rarely	Usually
Frequency	Rarely	Rarely	Usually
EMI	Rarely	Rarely	Usually
Other:			
Humidity	Occasionally	Occasionally	Usually
Dew Cycle	Rarely	Rarely	Occasionally
Altitude	Rarely	Rarely	Occasionally
Submersion	Rarely	Rarely	Occasionally
Dust	Rarely	Rarely	Occasionally
Wind/Rain	Rarely	Occasionally	Occasionally
Ultraviolet	Rarely	Rarely	Occasionally
Salt Spray	Rarely	Rarely	Occasionally
Salt Fog	Rarely	Rarely	Usually
Ozone	Rarely	Rarely	Occasionally

indicates that a specific environment is usually included as a phase of environmental testing for most products.

The actual environment that a product must survive varies greatly. For example, although vibration is rarely a concern for the storage phase, it is possible that an electronic box could be "stored" as a replacement unit on an ocean-going vessel, where it would be subjected to the vibration of heavy machinery near the storage area. Is the product to be stored in a shed without environmental control in the arctic or in the tropics? All aspects of the life of the product must be considered in detail, so that the definition of the real world that will be used for developing the laboratory test environments is accurate.

Products are usually stored in environmentally controlled buildings and within boxes or containers. Thus, they will rarely see thermal cycling or shock. However, when they are shipped (in their containers), they can be moved quickly from a thermally controlled environment to an extreme thermal environment, inducing thermal shock and cycling.

After the real world is defined, the next step in the process is to instrument the product when it is being subjected to the real world. For an environment such as thermal soak during storage, this probably is unnecessary. For an environment such as vibration under customer usage, a considerable amount of data must be acquired and reduced for developing the laboratory environments. The instrumentation, data acquisition, and data reduction processes require extensive efforts. Furthermore, for a given industry, the data must be reacquired on a regular basis to continually update the subsequent laboratory testing to ensure that it is current and representative.

Some environments, such as thermal shock, occur only occasionally in real-world situations but are commonly used for defect exposure testing. Specialized tests such as altitude and ozone depend on the anticipated application of the product by the customer. The severity of the test levels should be considered in developing a test program based on the definition of the real world. If the customer usage vibration levels greatly exceed those of the transportation environment, then it may be possible to exclude the transportation vibration environments from the test program.

The sequence of applied environments to be used in the testing program must be considered, based on the real world. The tests should follow the real-world sequence as closely as possible. If the facilities are available, then combined environmental testing should be performed. For example, vibration can often be combined with thermal cycling.

Stimulation

Stimulation testing is conducted in the absence of real-world data or to rapidly precipitate failures. The environment is applied based on predefined limitations for qualification testing. To precipitate failures, the applied levels are increased. Because it is conducted with arbitrary test levels, stimulation testing must be used with great care. Time and money are saved because the efforts of real-world definition, data acquisition, and data reduction are eliminated. However, the test results can be unreliable because they lack a solid basis in reality.

Application

Stimulation testing is performed to develop a product or to precipitate failures. A prototype product can be subjected to an environment either to survive a predetermined level or to determine its weaknesses by precipitating failures, which are then corrected. Also, field failures can possibly be duplicated rapidly in the absence of real-world data by applying laboratory environments.

Typical Test Types

The most commonly applied stimulation vibration test is a sinusoidal input. This can be a sweep of a frequency range or a dwell at a specific frequency. One procedure is to sweep a frequency range to determine the fundamental resonant frequency of the product, and then dwell at this frequency for a specified time at a defined level (g input). Thermal cycling and thermal shock are also stimulation tests that are commonly used. Other typical tests include a multiple shock-induced vibration combined with thermal cycling.

Types of Stresses

This section discusses the various types of stresses used in accelerated testing.

Temperature

Stresses—and failures—can be induced in hardware by thermal changes. These are usually due to differential thermal expansion. Assemblies are made at room temperature and then—in the real world—are subjected to environments from the arctic to the tropics. These assemblies are made of many different materials, each with a different coefficient of thermal expansion. When the temperature is changed, the different parts expand or contract different amounts. For electrical assemblies, breaks can occur in the circuitry. For structural assemblies, stresses are induced as two parts, which are attached to each other, to change the two different lengths.

Another cause of thermal-induced failures is the transition temperature of metals. Some metals and alloys exhibit the property that they become brittle and lose ductility when their temperature decreases to a certain value. Mechanical stresses applied to the material when it is below the transition temperature can induce catastrophic failures. A historic example of this is the Liberty ships that were produced during World War II. Many of these ships were made of steel, which had a marked transition temperature that was not known. The high seas, combined with cold winter temperatures, in the North Atlantic caused several of these ships to break in two and disappear without a trace.

Steady State

Steady-state thermal testing is used to evaluate the operating characteristics of hardware in environmental extremes.

- **Low Temperature**—Low-temperature steady-state testing is used for evaluating hardware in arctic conditions. Cycling of mechanical components (e.g., latches) at low temperature evaluates their durability characteristics. Cyclic load testing at steady-state low temperature exposes the brittleness of materials below their transition temperatures. Mechanical operation at low temperature is used for evaluating the effectiveness of lubricants, bushings, and bearings. Differential thermal expansion can cause seizing of moving parts. Elastomeric components can become brittle and crack. Bonding agents may become brittle and lose their adhesive capabilities.
- **High Temperature**—High-temperature steady-state testing is used for evaluating hardware in extreme tropical conditions. Mechanical operation at high temperature is used for evaluating the effectiveness of lubricants, bushings, and bearings. Elastomeric components

can lose pliability and become excessively soft, failing to perform their intended functions. Bonding agents may lose their adhesive capabilities.

Temperature Cycling

Temperature cycling is used to evaluate the hardware during exposure to thermal extremes. Due to differential expansion over several thermal cycles, it is possible to induce cracking. In electronic assemblies, this can be evidenced by open circuitry.

Thermal testing is performed within a chamber that has capabilities for extreme temperatures as well as rapid changes in temperature. Depending on the test being performed, there are two different approaches to fixturing. The test article can be set on an open framework so that all sides are exposed to the chamber ambient conditions. This maximizes the effects of the thermal cycling and induces maximum rate thermal changes within the test article. If the test article is attached to a large heat sink and the effects of the heat sink are critical to the operation of the unit at extreme temperature, then the heat sink should be simulated in the test. This is usually accomplished by mounting the test article on a large metallic block with the appropriate thermal mass.

The rate of thermal change in the chamber is often critical to inducing stresses that are potentially failure-inducing. Various methods can be incorporated into chambers to create rapid thermal changes. Standard heat-exchanger coolers can be augmented by the use of liquid nitrogen. Humidity also can be a problem in a thermal chamber. Condensation will form and freeze if ambient air is used in the chamber. The condensation problem can be handled by using a nitrogen atmosphere in the chamber.

During thermal cycle testing, it is possible to decrease the required soak time by overshooting the ambient temperature in the chamber. By exceeding the required soak temperature temporarily, the point of maximum thermal mass on the test article can reach the required soak temperature more quickly. When the point of maximum thermal mass reaches the required temperature, then the chamber ambient is brought back to this temperature for the soak. Care must be taken to ensure that the remainder of the test article can withstand the excessive temperature extreme without inducing failures.

Thermal Shock

Thermal shock is a test where the ambient temperature surrounding a test article is rapidly changed from one extreme to another. This is usually accomplished by thermally soaking the test article in a chamber at one temperature extreme and then quickly moving it to a second chamber at the other temperature extreme.

Dual thermal chambers are available, which can be used for thermal shock testing. The test article is placed in one chamber at one temperature extreme. After stabilizing at temperature, the test article is then automatically moved through a door in the wall between the two chambers and exposed to the opposite temperature extreme. The two chambers may contain various types of environments, such as the following:

- Air to air
- Air to liquid
- Liquid to liquid

Vibration

Vibration is possibly the most complex environment that is used for testing. In the early days of mass production and testing, only two options were available for testing: (1) expose the test article to the entire real-world environment, or (2) simulate it in the laboratory with a sinusoidal input. Although frequency and amplitude could be varied for the input, the real world is rarely sinusoidal, and these tests were at best a poor simulation. It required the advent of the computer so that vibration data could be reduced, understood, and reproduced in the laboratory.

Laboratory vibration testing falls into three primary categories:

1. Sinusoidal sweeps and dwells
2. Real-time simulation
3. Random vibration

Sinusoidal sweeps and dwells are used for evaluating resonant frequencies, mode shapes, and damping. They also can be used for quick hardware evaluation in the absence of any real-world data. Real-time simulation measures the output from the real world and uses it as an input for the laboratory test. Both sinusoidal vibration and real-time simulation are analyzed and reproduced in the "time domain." This means that the data are interpreted as a graph of amplitude versus time. Random vibration is analyzed and reproduced in the "frequency domain," meaning that the data are interpreted as a graph of amplitude versus frequency. This is an abstract concept that must be understood in order to effectively apply this means of testing. Random vibration can accurately simulate the complex vibration input experienced by hardware that is simultaneously exposed to several environments.

Other testing methods include mechanical vibration inputs, vibration induced by repetitive shocks, and shock input. Mechanical vibration is generally used for simulating unsecured cargo being transported in a rough environment. Vibration induced by repetitive shocks from multiple sources can be used to develop a random vibration input. Shock inputs simulate short-time-duration, high-frequency, high-level inputs to the product. All of these methods have advantages and disadvantages for use in laboratory testing.

The sine wave is the basis of all data analysis and reproduction. Combinations of sine waves of different amplitudes, durations, and frequencies are used to reproduce more complex environments.

Analysis Tools

There are two primary categories of analytical tools for working with vibration data: (1) tools for the time domain, and (2) tools for the frequency domain. The time domain tools are used for sinusoidal and real-time simulation testing. The primary time domain tool is the time history plot, which shows the variation of amplitude with respect to time. As the name implies, it displays the acceleration, displacement, and force history of a specific measurement. The frequency domain tools are used for random vibration testing. The primary tool in the frequency domain is the power spectral density (PSD) plot. Fourier transforms are used to define the energy level at each frequency of the measurement for developing the PSD. Correspondingly, a PSD plot is a concise method of defining a time history.

Mechanical and Pneumatic

Mechanical vibration falls into two distinct categories: (1) rotary inputs for loose cargo testing, and (2) cyclic inputs for testing mechanisms. Loose cargo testing is performed by placing a product in its shipping configuration onto a bed, which is then excited with a mechanical rotary input. Mechanisms such as latches, pedals, and door window cranks can be tested with pneumatic actuators to evaluate their capability to withstand multiple cyclic inputs.

Pneumatics operate using linear actuators. These are powered by "shop air" at approximately 100 psi. Some rotary actuators may be available. They can be controlled by computers or simple timed relays, which open and close the valves that duct the compressed air to the actuators. A primary advantage of pneumatic systems is that they are inexpensive and easy to use. Most laboratories and factories already have a supply of compressed air for running various tools. Actuators can be obtained that apply loads up to and greater than 1,000 pounds. The common range employed is usually less than 100 pounds.

Pneumatics are best for cyclic operational loads such as opening and closing doors and repeatedly operating mechanisms. They have an upper frequency limit of approximately 1 Hz. Pneumatic actuators are ineffective in applying vibration loads because their frequency range is very limited.

Loose Cargo Testing

The purpose of loose cargo testing is to determine if the product in its shipping container can survive the transportation environments applied to it. The frequency and amplitude of the input sinusoidal vibration can be varied. The product in its shipping container is placed onto the bed of a machine. The bed has a wooden floor and wooden walls. The bed is then activated by a rotary input for a specified length of time, allowing the product to move freely.

Cyclic Testing

The primary tool used for cyclic testing is pneumatic actuation. Pneumatic actuators are available in a wide range of displacement and load capabilities, in both linear and rotary configurations. The fixturing is specific to each setup. However, note that the fixturing must always be more durable than the hardware being tested. The test will usually require several sets of hardware to be evaluated, and it is cost effective to use the same fixturing for all of them.

Pneumatic systems are excellent for performing cyclic tests. Actuators can be obtained with many different stroke lengths and load ratings. However, pneumatic actuators usually are limited to frequencies below 1 Hz and correspondingly are ineffective for applying vibration environments.

Electrodynamic

An electrodynamic shaker is essentially a large loudspeaker coil. It is driven electrically by an amplifier. A computer is used to control it by applying sinusoidal or random vibration using an acceleration feedback. (A load cell for force feedback also can be used.) Electrodynamic shakers can be used for applying very-high-frequency inputs. The inputs usually are defined as random vibration PSD plots or sinusoidal waves that can be repeated easily on any setup without the requirement of drive file development.

Small components, engine-mounted components, brackets, fluid operating systems, and especially electronics can be tested using electrodynamic excitation, usually in one axis at a time. Electrodynamics are excellent for running modal surveys of structures, as well as determining the audible resonances for resolving noise problems.

The lower frequency range of electrodynamic shakers is limited by displacement. For old shakers, this is 1.0-in. double amplitude; for new shakers, it is 2.0-in. double amplitude. This usually translates to a lower frequency limit of approximately 5 to 10 Hz. The upper frequency limit of electrodynamic shakers is usually above 2000 Hz. Shakers are readily available in a load range from 50 force pounds up to and exceeding 35,000 force pounds.

Items tested using electrodynamics are mounted to a plate on or attached to the shaker head. A "slip table" can be attached to the shaker for obtaining axes of vibration other than those obtained on the shaker head. The fixture that attaches the test article to the shaker head or slip table must be rigid and free of all resonances in the frequency range of the test. The fixture is usually constructed of aluminum plate with a minimum thickness of 25 mm (1.0 in.). If more than a flat plate is required, then multiple plates are welded together using full penetration welds.

Electrodynamic shakers are excellent for testing electronics and are used for ESS of electronics assemblies. They can be used for applying real-world simulations such as servo-hydraulic systems; however, because of their bulkiness, they usually are limited to one axis of excitation at a time. Three-axis slip tables are available, but because of cross-talk problems, the three shakers driving them must be run open loop, and the bearings present control problems at high frequencies.

Hydraulic

Hydraulics operate using linear actuators. These are powered by hydraulic pumps and reservoirs at high pressure. Some rotary actuators are available. They are controlled by computers to apply sinusoidal or real-world simulation inputs using force, displacement, or acceleration feedbacks as controls, using servo-valves to control the flow of pressurized hydraulic fluid into the actuators.

Hydraulic systems can be used to apply high loads. Multiple actuators can be used to simultaneously apply loads in several directions. The primary application of hydraulics in dynamic testing is to apply real-world simulation of motor vehicles in multiple directions simultaneously. They are used extensively to simulate test tracks in the laboratory.

The lower frequency of hydraulic systems is limited by the available displacement on the actuators that are being used. Practically, this usually is approximately 2 Hz. The upper limit is approximately 40 Hz. Theoretically, higher frequencies are available, but they lack adequate controllability. Actuators are readily available that can apply loads up to approximately 10,000 pounds and higher.

Fixturing for servo-hydraulic testing can be of many different forms. For a simple (i.e., one actuator) test, a linkage is made between the actuator and the test article. For a "four-poster" test, four hydraulic actuators are mounted vertically with flat plate "wheel pans" placed on top of them. This is used to apply the vertical inputs that a vehicle experiences when on a rough surface. More complex setups can apply vertical, lateral, and torsional ("braking") loads at each wheel hub simultaneously, along with a torsional input due to the engine driveshaft.

Servo-hydraulics is good for simulating the loads on major structures. However, the higher frequencies that are present in the real world cannot be applied. Because of the 40-Hz limitation, any structures having resonances in excess of this frequency cannot be adequately tested. This includes all small components, brackets, and electronics. The "drive files" that are used for controlling multi-axis tests in real-world simulations are very setup specific. If the setup is changed in any way, a new drive file must be developed from the original data.

Mechanical Shock

Mechanical shock is defined as either a time history pulse or a shock response spectrum. The most commonly used time history pulses are defined as half sine waves or (triangular) sawtooths. Shock response spectra are defined as a plot of response accelerations (g) versus frequency (hertz). These accelerations represent the response that a single-degree-of-freedom system would experience at every frequency in the range when subjected to the required input. Shock response spectra are useful tools for analysis because if the resonant frequencies of a product are known, then the loads to which the product will be subjected are defined. A good reference for shock testing is MIL-STD-810, Section 516.

Mechanical shock can be induced by either a shock machine or an electrodynamic shaker. A shock machine has a platen upon which the test article is mounted. The platen is then raised a predetermined height and dropped on a crushable object. The configuration of the crushable object determines the resulting shock pulse. The shock machine is calibrated for the actual test by first conducting drops using dummy hardware and varying the object onto which the platen is dropped. When the desired shock pulse or shock response spectrum is obtained, then the actual hardware is mounted on the platen for the test. The fast response capability of an electrodynamic shaker also can be used for inducing a mechanical shock by programming the time history of the shock pulse into the shaker.

Fixturing for shock tests usually is minimal. The test article must be solidly mounted or clamped to the shock test machine in the desired orientation.

*Electrical Stress**

Electrical stress is used to test a product near or at its electrical limits. Exercising circuitry and simulating junction temperatures on semiconductors are both good examples of electrical stress tests. There are two basic types of electrical stress tests: (1) power cycling, and (2) voltage margining.

Power cycling consists of turning product power on and off at specified levels. Voltage margining involves varying input power above and below nominal product power requirements. A subset of voltage margining is frequency margining.

Typically, electrical stress alone does not expose the number of defects commonly found through thermal cycling and vibration stress. However, because it is typically necessary to supply power to products to find the soft or intermittent failures, it can be relatively inexpensive to implement electrical stress in combination with other stress environments to increase overall effectiveness.

Combined Environments

Although each environment (e.g., vibration, thermal cycling) must be separated and understood individually, in the real world, these environments are combined. A car or truck traverses rough roads in both the summer and winter, in the tropics and in the arctic, at sea level and high in the mountains. The most representative laboratory tests can combine the vibration, shock, and thermal environments, as well as others.

A common combination of environments is that of vibration and thermal cycling. Thermal chambers are made for placing over electrodynamic shakers. Using these, it is possible to simultaneously apply a vibration input combined with thermal extremes and thermal cycles. For example, it is possible that an electrical connection could be cracked due to thermal extremes and differential thermal expansion. However, because the two halves of the joint are in contact, the electrical performance is unaffected. When a vibration environment is applied, the resulting motion causes physical separation of the halves of their joint, resulting in an intermittent short in the electrical circuit and exposing the defect. On a larger scale, it is possible to construct a test setup where an entire car or truck can be mounted for multiple axes of real-time simulation vibration input in a chamber where thermal extremes can be applied.

Other Types of Stress

In addition to the standard thermal, humidity, vibration, and electrical environments that are tested, many other real-world environments must be considered. Depending on the location of

---

* This material is used courtesy of Thermotron Industries, www.thermotron.com.

a component and its anticipated usage, it could be adversely affected by many different factors. Salt fog and salt spray can cause corrosion and can accelerate the effects of dissimilar metals. Ozone and ultraviolet light can cause deterioration of elastomers and coatings. Moisture and dust driven by wind can penetrate and damage many devices. All of these must be considered when developing a laboratory test.

1. **Altitude (MIL-STD-810, Method No. 500.3)**—The low pressure of high altitude can cause pressure loss within volumes that are nominally sealed. When the unit returns to sea level, it then can ingest air with impurities, which can adversely affect the unit.

2. **Submersion (MIL-STD-810, Method No. 512.3)**—Submersion is a test usually applied to products that are applied to marine environments where the possibility exists that the products could become completely submerged in water. The submersion test evaluates the effectiveness of the seals in the product.

3. **Dust (MIL-STD-810, Method No. 510.3)**—Dust, sand, and other impurities can be detrimental to a product. These particles can be extremely abrasive when driven by wind.

4. **Moisture, Wind, and Rain (MIL-STD-810, Method No. 506.3)**—Moisture driven by wind can penetrate seals and cause damage within a product.

5. **Ultraviolet (MIL-STD-810, Method No. 505.3)**—Ultraviolet radiation can cause heating of the product in a different manner than that of a thermal soak.

6. **Corrosion**—Several forms of corrosion can affect the performance of a product. These include salt spray and salt fog (MIL-STD-810, Method No. 509.3), brake fluid, soft drinks, and lotions.

7. **Explosive Atmosphere (MIL-STD-810, Method No. 511.3)**—Explosive atmospheric testing is used to determine if the electronics in a product can ignite a potentially explosive atmosphere in which it is operated. This primarily is applicable to fuel system components.

8. **Fungus (MIL-STD-810, Method No. 508.4)**—Fungus infiltration can be detrimental to the durability and operation of a product.

9. **Icing (MIL-STD-810, Method No. 508.4)**—Icing and freezing rain can adversely affect the operation of components that are exposed to these environments.

Summary

Stress environments can be used individually or in combination. The duration of the tests containing these stresses may be measured in seconds or months. In general, short-duration tests targeted to specific failure modes are used during product development. This type of test is likely to use a single type of stress. Qualification tests are typically longer in duration and use multiple types of stress.

Appendix A

Statistical Tables

TABLE A.1
THE GAMMA FUNCTION

x	G(**x**)	x	G(**x**)	x	G(**x**)	x	G(**x**)
1.00	1.0000	1.25	0.9064	1.50	0.8862	1.75	0.9191
1.01	0.9943	1.26	0.9044	1.51	0.8866	1.76	0.9214
1.02	0.9888	1.27	0.9025	1.52	0.8870	1.77	0.9238
1.03	0.9835	1.28	0.9007	1.53	0.8876	1.78	0.9262
1.04	0.9784	1.29	0.8990	1.54	0.8882	1.79	0.9288
1.05	0.9735	1.30	0.8975	1.55	0.8889	1.80	0.9314
1.06	0.9687	1.31	0.8960	1.56	0.8896	1.81	0.9341
1.07	0.9642	1.32	0.8946	1.57	0.8905	1.82	0.9368
1.08	0.9597	1.33	0.8934	1.58	0.8914	1.83	0.9397
1.09	0.9555	1.34	0.8922	1.59	0.8924	1.84	0.9426
1.10	0.9514	1.35	0.8912	1.60	0.8935	1.85	0.9456
1.11	0.9474	1.36	0.8902	1.61	0.8947	1.86	0.9487
1.12	0.9436	1.37	0.8893	1.62	0.8959	1.87	0.9518
1.13	0.9399	1.38	0.8885	1.63	0.8972	1.88	0.9551
1.14	0.9364	1.39	0.8879	1.64	0.8986	1.89	0.9584
1.15	0.9330	1.40	0.8873	1.65	0.9001	1.90	0.9618
1.16	0.9298	1.41	0.8868	1.66	0.9017	1.91	0.9652
1.17	0.9267	1.42	0.8864	1.67	0.9033	1.92	0.9688
1.18	0.9237	1.43	0.8860	1.68	0.9050	1.93	0.9724
1.19	0.9209	1.44	0.8858	1.69	0.9068	1.94	0.9761
1.20	0.9182	1.45	0.8857	1.70	0.9086	1.95	0.9799
1.21	0.9156	1.46	0.8856	1.71	0.9106	1.96	0.9837
1.22	0.9131	1.47	0.8856	1.72	0.9126	1.97	0.9877
1.23	0.9108	1.48	0.8857	1.73	0.9147	1.98	0.9917
1.24	0.9085	1.49	0.8859	1.74	0.9168	1.99	0.9958
						2.00	1.0000

$$\Gamma(x) = \int_0^\infty e^{-\tau} \tau^{(x-1)} d\tau$$

$$\Gamma(x+1) = x\Gamma(x)$$

TABLE A.2
STANDARD NORMAL CUMULATIVE DISTRIBUTION FUNCTION

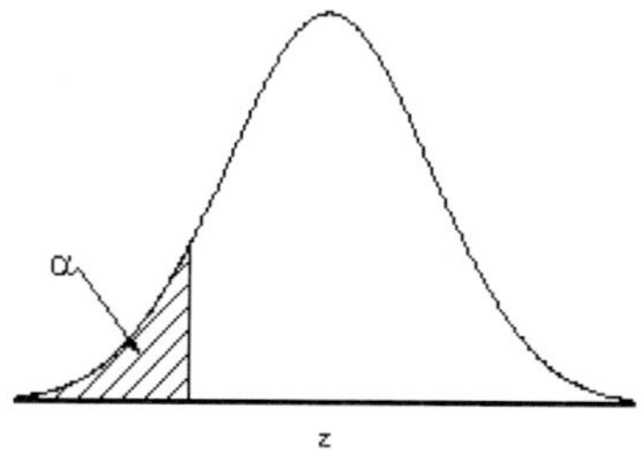

z	0.00	0.01	0.02	0.03	0.04	0.05	0.06	0.07	0.08	0.09
-3.40	0.0003	0.0003	0.0004	0.0004	0.0004	0.0004	0.0004	0.0004	0.0005	0.0005
-3.30	0.0005	0.0005	0.0005	0.0005	0.0006	0.0006	0.0006	0.0006	0.0006	0.0007
-3.20	0.0007	0.0007	0.0007	0.0008	0.0008	0.0008	0.0008	0.0009	0.0009	0.0009
-3.10	0.0010	0.0010	0.0010	0.0011	0.0011	0.0011	0.0012	0.0012	0.0013	0.0013
-3.00	0.0013	0.0014	0.0014	0.0015	0.0015	0.0016	0.0016	0.0017	0.0018	0.0018
-2.90	0.0019	0.0019	0.0020	0.0021	0.0021	0.0022	0.0023	0.0023	0.0024	0.0025
-2.80	0.0026	0.0026	0.0027	0.0028	0.0029	0.0030	0.0031	0.0032	0.0033	0.0034
-2.70	0.0035	0.0036	0.0037	0.0038	0.0039	0.0040	0.0041	0.0043	0.0044	0.0045
-2.60	0.0047	0.0048	0.0049	0.0051	0.0052	0.0054	0.0055	0.0057	0.0059	0.0060
-2.50	0.0062	0.0064	0.0066	0.0068	0.0069	0.0071	0.0073	0.0075	0.0078	0.0080
-2.40	0.0082	0.0084	0.0087	0.0089	0.0091	0.0094	0.0096	0.0099	0.0102	0.0104
-2.30	0.0107	0.0110	0.0113	0.0116	0.0119	0.0122	0.0125	0.0129	0.0132	0.0136
-2.20	0.0139	0.0143	0.0146	0.0150	0.0154	0.0158	0.0162	0.0166	0.0170	0.0174
-2.10	0.0179	0.0183	0.0188	0.0192	0.0197	0.0202	0.0207	0.0212	0.0217	0.0222
-2.00	0.0228	0.0233	0.0239	0.0244	0.0250	0.0256	0.0262	0.0268	0.0274	0.0281
-1.90	0.0287	0.0294	0.0301	0.0307	0.0314	0.0322	0.0329	0.0336	0.0344	0.0351
-1.80	0.0359	0.0367	0.0375	0.0384	0.0392	0.0401	0.0409	0.0418	0.0427	0.0436
-1.70	0.0446	0.0455	0.0465	0.0475	0.0485	0.0495	0.0505	0.0516	0.0526	0.0537
-1.60	0.0548	0.0559	0.0571	0.0582	0.0594	0.0606	0.0618	0.0630	0.0643	0.0655
-1.50	0.0668	0.0681	0.0694	0.0708	0.0721	0.0735	0.0749	0.0764	0.0778	0.0793
-1.40	0.0808	0.0823	0.0838	0.0853	0.0869	0.0885	0.0901	0.0918	0.0934	0.0951
-1.30	0.0968	0.0985	0.1003	0.1020	0.1038	0.1056	0.1075	0.1093	0.1112	0.1131
-1.20	0.1151	0.1170	0.1190	0.1210	0.1230	0.1251	0.1271	0.1292	0.1314	0.1335
-1.10	0.1357	0.1379	0.1401	0.1423	0.1446	0.1469	0.1492	0.1515	0.1539	0.1562
-1.00	0.1587	0.1611	0.1635	0.1660	0.1685	0.1711	0.1736	0.1762	0.1788	0.1814
-0.90	0.1841	0.1867	0.1894	0.1922	0.1949	0.1977	0.2005	0.2033	0.2061	0.2090
-0.80	0.2119	0.2148	0.2177	0.2206	0.2236	0.2266	0.2296	0.2327	0.2358	0.2389

TABLE A.2
STANDARD NORMAL CUMULATIVE DISTRIBUTION FUNCTION *(cont.)*

z	0.00	0.01	0.02	0.03	0.04	0.05	0.06	0.07	0.08	0.09
-0.70	0.2420	0.2451	0.2483	0.2514	0.2546	0.2578	0.2611	0.2643	0.2676	0.2709
-0.60	0.2743	0.2776	0.2810	0.2843	0.2877	0.2912	0.2946	0.2981	0.3015	0.3050
-0.50	0.3085	0.3121	0.3156	0.3192	0.3228	0.3264	0.3300	0.3336	0.3372	0.3409
-0.40	0.3446	0.3483	0.3520	0.3557	0.3594	0.3632	0.3669	0.3707	0.3745	0.3783
-0.30	0.3821	0.3859	0.3897	0.3936	0.3974	0.4013	0.4052	0.4090	0.4129	0.4168
-0.20	0.4207	0.4247	0.4286	0.4325	0.4364	0.4404	0.4443	0.4483	0.4522	0.4562
-0.10	0.4602	0.4641	0.4681	0.4721	0.4761	0.4801	0.4840	0.4880	0.4920	0.4960
0.00	0.5000	0.5040	0.5080	0.5120	0.5160	0.5199	0.5239	0.5279	0.5319	0.5359
0.10	0.5398	0.5438	0.5478	0.5517	0.5557	0.5596	0.5636	0.5675	0.5714	0.5753
0.20	0.5793	0.5832	0.5871	0.5910	0.5948	0.5987	0.6026	0.6064	0.6103	0.6141
0.30	0.6179	0.6217	0.6255	0.6293	0.6331	0.6368	0.6406	0.6443	0.6480	0.6517
0.40	0.6554	0.6591	0.6628	0.6664	0.6700	0.6736	0.6772	0.6808	0.6844	0.6879
0.50	0.6915	0.6950	0.6985	0.7019	0.7054	0.7088	0.7123	0.7157	0.7190	0.7224
0.60	0.7257	0.7291	0.7324	0.7357	0.7389	0.7422	0.7454	0.7486	0.7517	0.7549
0.70	0.7580	0.7611	0.7642	0.7673	0.7704	0.7734	0.7764	0.7794	0.7823	0.7852
0.80	0.7881	0.7910	0.7939	0.7967	0.7995	0.8023	0.8051	0.8078	0.8106	0.8133
0.90	0.8159	0.8186	0.8212	0.8238	0.8264	0.8289	0.8315	0.8340	0.8365	0.8389
1.00	0.8413	0.8438	0.8461	0.8485	0.8508	0.8531	0.8554	0.8577	0.8599	0.8621
1.10	0.8643	0.8665	0.8686	0.8708	0.8729	0.8749	0.8770	0.8790	0.8810	0.8830
1.20	0.8849	0.8869	0.8888	0.8907	0.8925	0.8944	0.8962	0.8980	0.8997	0.9015
1.30	0.9032	0.9049	0.9066	0.9082	0.9099	0.9115	0.9131	0.9147	0.9162	0.9177
1.40	0.9192	0.9207	0.9222	0.9236	0.9251	0.9265	0.9279	0.9292	0.9306	0.9319
1.50	0.9332	0.9345	0.9357	0.9370	0.9382	0.9394	0.9406	0.9418	0.9429	0.9441
1.60	0.9452	0.9463	0.9474	0.9484	0.9495	0.9505	0.9515	0.9525	0.9535	0.9545
1.70	0.9554	0.9564	0.9573	0.9582	0.9591	0.9599	0.9608	0.9616	0.9625	0.9633
1.80	0.9641	0.9649	0.9656	0.9664	0.9671	0.9678	0.9686	0.9693	0.9699	0.9706
1.90	0.9713	0.9719	0.9726	0.9732	0.9738	0.9744	0.9750	0.9756	0.9761	0.9767
2.00	0.9772	0.9778	0.9783	0.9788	0.9793	0.9798	0.9803	0.9808	0.9812	0.9817
2.10	0.9821	0.9826	0.9830	0.9834	0.9838	0.9842	0.9846	0.9850	0.9854	0.9857
2.20	0.9861	0.9864	0.9868	0.9871	0.9875	0.9878	0.9881	0.9884	0.9887	0.9890
2.30	0.9893	0.9896	0.9898	0.9901	0.9904	0.9906	0.9909	0.9911	0.9913	0.9916
2.40	0.9918	0.9920	0.9922	0.9925	0.9927	0.9929	0.9931	0.9932	0.9934	0.9936
2.50	0.9938	0.9940	0.9941	0.9943	0.9945	0.9946	0.9948	0.9949	0.9951	0.9952
2.60	0.9953	0.9955	0.9956	0.9957	0.9959	0.9960	0.9961	0.9962	0.9963	0.9964
2.70	0.9965	0.9966	0.9967	0.9968	0.9969	0.9970	0.9971	0.9972	0.9973	0.9974
2.80	0.9974	0.9975	0.9976	0.9977	0.9977	0.9978	0.9979	0.9979	0.9980	0.9981
2.90	0.9981	0.9982	0.9982	0.9983	0.9984	0.9984	0.9985	0.9985	0.9986	0.9986
3.00	0.9987	0.9987	0.9987	0.9988	0.9988	0.9989	0.9989	0.9989	0.9990	0.9990
3.10	0.9990	0.9991	0.9991	0.9991	0.9992	0.9992	0.9992	0.9992	0.9993	0.9993
3.20	0.9993	0.9993	0.9994	0.9994	0.9994	0.9994	0.9994	0.9995	0.9995	0.9995
3.30	0.9995	0.9995	0.9995	0.9996	0.9996	0.9996	0.9996	0.9996	0.9996	0.9997
3.40	0.9997	0.9997	0.9997	0.9997	0.9997	0.9997	0.9997	0.9997	0.9997	0.9998

TABLE A.3
CHI-SQUARE SIGNIFICANCE

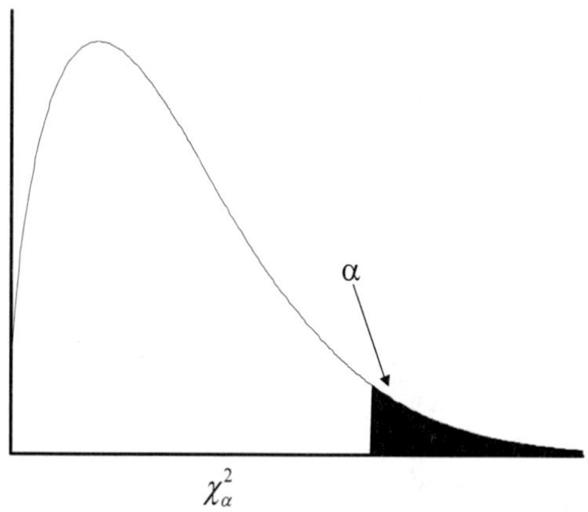

	α								
γ	0.005	0.01	0.02	0.025	0.05	0.1	0.2	0.25	0.3
1	7.8794	6.6349	5.4119	5.0239	3.8415	2.7055	1.6424	1.3233	1.0742
2	10.5966	9.2103	7.8240	7.3778	5.9915	4.6052	3.2189	2.7726	2.4079
3	12.8382	11.3449	9.8374	9.3484	7.8147	6.2514	4.6416	4.1083	3.6649
4	14.8603	13.2767	11.6678	11.1433	9.4877	7.7794	5.9886	5.3853	4.8784
5	16.7496	15.0863	13.3882	12.8325	11.0705	9.2364	7.2893	6.6257	6.0644
6	18.5476	16.8119	15.0332	14.4494	12.5916	10.6446	8.5581	7.8408	7.2311
7	20.2777	18.4753	16.6224	16.0128	14.0671	12.0170	9.8032	9.0371	8.3834
8	21.9550	20.0902	18.1682	17.5345	15.5073	13.3616	11.0301	10.2189	9.5245
9	23.5894	21.6660	19.6790	19.0228	16.9190	14.6837	12.2421	11.3888	10.6564
10	25.1882	23.2093	21.1608	20.4832	18.3070	15.9872	13.4420	12.5489	11.7807
11	26.7568	24.7250	22.6179	21.9200	19.6751	17.2750	14.6314	13.7007	12.8987
12	28.2995	26.2170	24.0540	23.3367	21.0261	18.5493	15.8120	14.8454	14.0111
13	29.8195	27.6882	25.4715	24.7356	22.3620	19.8119	16.9848	15.9839	15.1187
14	31.3193	29.1412	26.8728	26.1189	23.6848	21.0641	18.1508	17.1169	16.2221
15	32.8013	30.5779	28.2595	27.4884	24.9958	22.3071	19.3107	18.2451	17.3217
16	34.2672	31.9999	29.6332	28.8454	26.2962	23.5418	20.4651	19.3689	18.4179
17	35.7185	33.4087	30.9950	30.1910	27.5871	24.7690	21.6146	20.4887	19.5110
18	37.1565	34.8053	32.3462	31.5264	28.8693	25.9894	22.7595	21.6049	20.6014
19	38.5823	36.1909	33.6874	32.8523	30.1435	27.2036	23.9004	22.7178	21.6891
20	39.9968	37.5662	35.0196	34.1696	31.4104	28.4120	25.0375	23.8277	22.7745
21	41.4011	38.9322	36.3434	35.4789	32.6706	29.6151	26.1711	24.9348	23.8578
22	42.7957	40.2894	37.6595	36.7807	33.9244	30.8133	27.3015	26.0393	24.9390
23	44.1813	41.6384	38.9683	38.0756	35.1725	32.0069	28.4288	27.1413	26.0184
24	45.5585	42.9798	40.2704	39.3641	36.4150	33.1962	29.5533	28.2412	27.0960
25	46.9279	44.3141	41.5661	40.6465	37.6525	34.3816	30.6752	29.3389	28.1719
26	48.2899	45.6417	42.8558	41.9232	38.8851	35.5632	31.7946	30.4346	29.2463
27	49.6449	46.9629	44.1400	43.1945	40.1133	36.7412	32.9117	31.5284	30.3193
28	50.9934	48.2782	45.4188	44.4608	41.3371	37.9159	34.0266	32.6205	31.3909
29	52.3356	49.5879	46.6927	45.7223	42.5570	39.0875	35.1394	33.7109	32.4612
30	53.6720	50.8922	47.9618	46.9792	43.7730	40.2560	36.2502	34.7997	33.5302

TABLE A.3
CHI-SQUARE SIGNIFICANCE *(cont.)*

α									
0.5	0.6	0.7	0.8	0.85	0.9	0.95	0.975	0.99	0.995
0.4549	0.2750	0.1485	0.0642	0.0358	0.0158	0.0039	0.0010	0.0002	0.0000
1.3863	1.0217	0.7133	0.4463	0.3250	0.2107	0.1026	0.0506	0.0201	0.0100
2.3660	1.8692	1.4237	1.0052	0.7978	0.5844	0.3518	0.2158	0.1148	0.0717
3.3567	2.7528	2.1947	1.6488	1.3665	1.0636	0.7107	0.4844	0.2971	0.2070
4.3515	3.6555	2.9999	2.3425	1.9938	1.6103	1.1455	0.8312	0.5543	0.4117
5.3481	4.5702	3.8276	3.0701	2.6613	2.2041	1.6354	1.2373	0.8721	0.6757
6.3458	5.4932	4.6713	3.8223	3.3583	2.8331	2.1673	1.6899	1.2390	0.9893
7.3441	6.4226	5.5274	4.5936	4.0782	3.4895	2.7326	2.1797	1.6465	1.3444
8.3428	7.3570	6.3933	5.3801	4.8165	4.1682	3.3251	2.7004	2.0879	1.7349
9.3418	8.2955	7.2672	6.1791	5.5701	4.8652	3.9403	3.2470	2.5582	2.1559
10.3410	9.2373	8.1479	6.9887	6.3364	5.5778	4.5748	3.8157	3.0535	2.6032
11.3403	10.1820	9.0343	7.8073	7.1138	6.3038	5.2260	4.4038	3.5706	3.0738
12.3398	11.1291	9.9257	8.6339	7.9008	7.0415	5.8919	5.0088	4.1069	3.5650
13.3393	12.0785	10.8215	9.4673	8.6963	7.7895	6.5706	5.6287	4.6604	4.0747
14.3389	13.0297	11.7212	10.3070	9.4993	8.5468	7.2609	6.2621	5.2293	4.6009
15.3385	13.9827	12.6243	11.1521	10.3090	9.3122	7.9616	6.9077	5.8122	5.1422
16.3382	14.9373	13.5307	12.0023	11.1249	10.0852	8.6718	7.5642	6.4078	5.6972
17.3379	15.8932	14.4399	12.8570	11.9463	10.8649	9.3905	8.2307	7.0149	6.2648
18.3377	16.8504	15.3517	13.7158	12.7727	11.6509	10.1170	8.9065	7.6327	6.8440
19.3374	17.8088	16.2659	14.5784	13.6039	12.4426	10.8508	9.5908	8.2604	7.4338
20.3372	18.7683	17.1823	15.4446	14.4393	13.2396	11.5913	10.2829	8.8972	8.0337
21.3370	19.7288	18.1007	16.3140	15.2788	14.0415	12.3380	10.9823	9.5425	8.6427
22.3369	20.6902	19.0211	17.1865	16.1219	14.8480	13.0905	11.6886	10.1957	9.2604
23.3367	21.6525	19.9432	18.0618	16.9686	15.6587	13.8484	12.4012	10.8564	9.8862
24.3366	22.6156	20.8670	18.9398	17.8184	16.4734	14.6114	13.1197	11.5240	10.5197
25.3365	23.5794	21.7924	19.8202	18.6714	17.2919	15.3792	13.8439	12.1981	11.1602
26.3363	24.5440	22.7192	20.7030	19.5272	18.1139	16.1514	14.5734	12.8785	11.8076
27.3362	25.5093	23.6475	21.5880	20.3857	18.9392	16.9279	15.3079	13.5647	12.4613
28.3361	26.4751	24.5770	22.4751	21.2468	19.7677	17.7084	16.0471	14.2565	13.1211
29.3360	27.4416	25.5078	23.3641	22.1103	20.5992	18.4927	16.7908	14.9535	13.7867

TABLE A.4
F SIGNIFICANCE

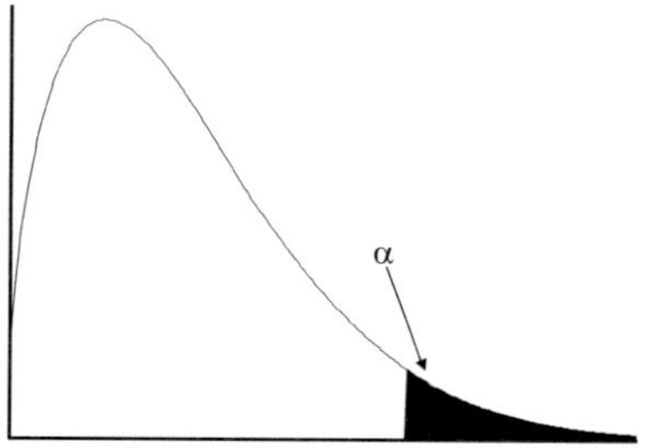

TABLE A.4.1 α = 0.01

γ1 \ γ2	1	2	3	4	5	6	7	8	9	10	15	20	25	50	100
1	4052.18	98.50	34.12	21.20	16.26	13.75	12.25	11.26	10.56	10.04	8.68	8.10	7.77	7.17	6.90
2	4999.50	99.00	30.82	18.00	13.27	10.92	9.55	8.65	8.02	7.56	6.36	5.85	5.57	5.06	4.82
3	5403.35	99.17	29.46	16.69	12.06	9.78	8.45	7.59	6.99	6.55	5.42	4.94	4.68	4.20	3.98
4	5624.58	99.25	28.71	15.98	11.39	9.15	7.85	7.01	6.42	5.99	4.89	4.43	4.18	3.72	3.51
5	5763.65	99.30	28.24	15.52	10.97	8.75	7.46	6.63	6.06	5.64	4.56	4.10	3.85	3.41	3.21
6	5858.99	99.33	27.91	15.21	10.67	8.47	7.19	6.37	5.80	5.39	4.32	3.87	3.63	3.19	2.99
7	5928.36	99.36	27.67	14.98	10.46	8.26	6.99	6.18	5.61	5.20	4.14	3.70	3.46	3.02	2.82
8	5981.07	99.37	27.49	14.80	10.29	8.10	6.84	6.03	5.47	5.06	4.00	3.56	3.32	2.89	2.69
9	6022.47	99.39	27.35	14.66	10.16	7.98	6.72	5.91	5.35	4.94	3.89	3.46	3.22	2.78	2.59
10	6055.85	99.40	27.23	14.55	10.05	7.87	6.62	5.81	5.26	4.85	3.80	3.37	3.13	2.70	2.50
15	6157.28	99.43	26.87	14.20	9.72	7.56	6.31	5.52	4.96	4.56	3.52	3.09	2.85	2.42	2.22
20	6208.73	99.45	26.69	14.02	9.55	7.40	6.16	5.36	4.81	4.41	3.37	2.94	2.70	2.27	2.07
25	6239.83	99.46	26.58	13.91	9.45	7.30	6.06	5.26	4.71	4.31	3.28	2.84	2.60	2.17	1.97
50	6302.52	99.48	26.35	13.69	9.24	7.09	5.86	5.07	4.52	4.12	3.08	2.64	2.40	1.95	1.74
100	6334.11	99.49	26.24	13.58	9.13	6.99	5.75	4.96	4.41	4.01	2.98	2.54	2.29	1.82	1.60

TABLE A.4.2 α = 0.025

γ1 \ γ2	1	2	3	4	5	6	7	8	9	10	15	20	25	50	100
1	647.79	38.51	17.44	12.22	10.01	8.81	8.07	7.57	7.21	6.94	6.20	5.87	5.69	5.34	5.18
2	799.50	39.00	16.04	10.65	8.43	7.26	6.54	6.06	5.71	5.46	4.77	4.46	4.29	3.97	3.83
3	864.16	39.17	15.44	9.98	7.76	6.60	5.89	5.42	5.08	4.83	4.15	3.86	3.69	3.39	3.25
4	899.58	39.25	15.10	9.60	7.39	6.23	5.52	5.05	4.72	4.47	3.80	3.51	3.35	3.05	2.92
5	921.85	39.30	14.88	9.36	7.15	5.99	5.29	4.82	4.48	4.24	3.58	3.29	3.13	2.83	2.70
6	937.11	39.33	14.73	9.20	6.98	5.82	5.12	4.65	4.32	4.07	3.41	3.13	2.97	2.67	2.54
7	948.22	39.36	14.62	9.07	6.85	5.70	4.99	4.53	4.20	3.95	3.29	3.01	2.85	2.55	2.42
8	956.66	39.37	14.54	8.98	6.76	5.60	4.90	4.43	4.10	3.85	3.20	2.91	2.75	2.46	2.32
9	963.28	39.39	14.47	8.90	6.68	5.52	4.82	4.36	4.03	3.78	3.12	2.84	2.68	2.38	2.24
10	968.63	39.40	14.42	8.84	6.62	5.46	4.76	4.30	3.96	3.72	3.06	2.77	2.61	2.32	2.18
15	984.87	39.43	14.25	8.66	6.43	5.27	4.57	4.10	3.77	3.52	2.86	2.57	2.41	2.11	1.97
20	993.10	39.45	14.17	8.56	6.33	5.17	4.47	4.00	3.67	3.42	2.76	2.46	2.30	1.99	1.85
25	998.08	39.46	14.12	8.50	6.27	5.11	4.40	3.94	3.60	3.35	2.69	2.40	2.23	1.92	1.77
50	1008.12	39.48	14.01	8.38	6.14	4.98	4.28	3.81	3.47	3.22	2.55	2.25	2.08	1.75	1.59
100	1013.17	39.49	13.96	8.32	6.08	4.92	4.21	3.74	3.40	3.15	2.47	2.17	2.00	1.66	1.48

TABLE A.4
F SIGNIFICANCE *(cont.)*

TABLE A.4.3 $\alpha = 0.05$

	γ2														
γ1	1	2	3	4	5	6	7	8	9	10	15	20	25	50	100
1	161.45	18.51	10.13	7.71	6.61	5.99	5.59	5.32	5.12	4.96	4.54	4.35	4.24	4.03	3.94
2	199.50	19.00	9.55	6.94	5.79	5.14	4.74	4.46	4.26	4.10	3.68	3.49	3.39	3.18	3.09
3	215.71	19.16	9.28	6.59	5.41	4.76	4.35	4.07	3.86	3.71	3.29	3.10	2.99	2.79	2.70
4	224.58	19.25	9.12	6.39	5.19	4.53	4.12	3.84	3.63	3.48	3.06	2.87	2.76	2.56	2.46
5	230.16	19.30	9.01	6.26	5.05	4.39	3.97	3.69	3.48	3.33	2.90	2.71	2.60	2.40	2.31
6	233.99	19.33	8.94	6.16	4.95	4.28	3.87	3.58	3.37	3.22	2.79	2.60	2.49	2.29	2.19
7	236.77	19.35	8.89	6.09	4.88	4.21	3.79	3.50	3.29	3.14	2.71	2.51	2.40	2.20	2.10
8	238.88	19.37	8.85	6.04	4.82	4.15	3.73	3.44	3.23	3.07	2.64	2.45	2.34	2.13	2.03
9	240.54	19.38	8.81	6.00	4.77	4.10	3.68	3.39	3.18	3.02	2.59	2.39	2.28	2.07	1.97
10	241.88	19.40	8.79	5.96	4.74	4.06	3.64	3.35	3.14	2.98	2.54	2.35	2.24	2.03	1.93
15	245.95	19.43	8.70	5.86	4.62	3.94	3.51	3.22	3.01	2.85	2.40	2.20	2.09	1.87	1.77
20	248.01	19.45	8.66	5.80	4.56	3.87	3.44	3.15	2.94	2.77	2.33	2.12	2.01	1.78	1.68
25	249.26	19.46	8.63	5.77	4.52	3.83	3.40	3.11	2.89	2.73	2.28	2.07	1.96	1.73	1.62
50	251.77	19.48	8.58	5.70	4.44	3.75	3.32	3.02	2.80	2.64	2.18	1.97	1.84	1.60	1.48
100	253.04	19.49	8.55	5.66	4.41	3.71	3.27	2.97	2.76	2.59	2.12	1.91	1.78	1.52	1.39

TABLE A.4.4 $\alpha = 0.1$

	γ2														
γ1	1	2	3	4	5	6	7	8	9	10	15	20	25	50	100
1	39.86	8.53	5.54	4.54	4.06	3.78	3.59	3.46	3.36	3.29	3.07	2.97	2.92	2.81	2.76
2	49.50	9.00	5.46	4.32	3.78	3.46	3.26	3.11	3.01	2.92	2.70	2.59	2.53	2.41	2.36
3	53.59	9.16	5.39	4.19	3.62	3.29	3.07	2.92	2.81	2.73	2.49	2.38	2.32	2.20	2.14
4	55.83	9.24	5.34	4.11	3.52	3.18	2.96	2.81	2.69	2.61	2.36	2.25	2.18	2.06	2.00
5	57.24	9.29	5.31	4.05	3.45	3.11	2.88	2.73	2.61	2.52	2.27	2.16	2.09	1.97	1.91
6	58.20	9.33	5.28	4.01	3.40	3.05	2.83	2.67	2.55	2.46	2.21	2.09	2.02	1.90	1.83
7	58.91	9.35	5.27	3.98	3.37	3.01	2.78	2.62	2.51	2.41	2.16	2.04	1.97	1.84	1.78
8	59.44	9.37	5.25	3.95	3.34	2.98	2.75	2.59	2.47	2.38	2.12	2.00	1.93	1.80	1.73
9	59.86	9.38	5.24	3.94	3.32	2.96	2.72	2.56	2.44	2.35	2.09	1.96	1.89	1.76	1.69
10	60.19	9.39	5.23	3.92	3.30	2.94	2.70	2.54	2.42	2.32	2.06	1.94	1.87	1.73	1.66
15	61.22	9.42	5.20	3.87	3.24	2.87	2.63	2.46	2.34	2.24	1.97	1.84	1.77	1.63	1.56
20	61.74	9.44	5.18	3.84	3.21	2.84	2.59	2.42	2.30	2.20	1.92	1.79	1.72	1.57	1.49
25	62.05	9.45	5.17	3.83	3.19	2.81	2.57	2.40	2.27	2.17	1.89	1.76	1.68	1.53	1.45
50	62.69	9.47	5.15	3.80	3.15	2.77	2.52	2.35	2.22	2.12	1.83	1.69	1.61	1.44	1.35
100	63.01	9.48	5.14	3.78	3.13	2.75	2.50	2.32	2.19	2.09	1.79	1.65	1.56	1.39	1.29

TABLE A.5
***t* SIGNIFICANCE**

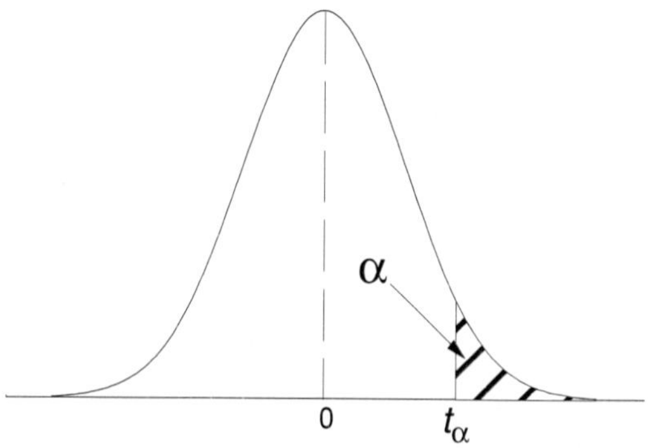

γ	0.1	0.05	0.025	0.01	0.005
1	3.0777	6.3138	12.7062	31.8205	63.6567
2	1.8856	2.9200	4.3027	6.9646	9.9248
3	1.6377	2.3534	3.1824	4.5407	5.8409
4	1.5332	2.1318	2.7764	3.7469	4.6041
5	1.4759	2.0150	2.5706	3.3649	4.0321
6	1.4398	1.9432	2.4469	3.1427	3.7074
7	1.4149	1.8946	2.3646	2.9980	3.4995
8	1.3968	1.8595	2.3060	2.8965	3.3554
9	1.3830	1.8331	2.2622	2.8214	3.2498
10	1.3722	1.8125	2.2281	2.7638	3.1693
11	1.3634	1.7959	2.2010	2.7181	3.1058
12	1.3562	1.7823	2.1788	2.6810	3.0545
13	1.3502	1.7709	2.1604	2.6503	3.0123
14	1.3450	1.7613	2.1448	2.6245	2.9768
15	1.3406	1.7531	2.1314	2.6025	2.9467
16	1.3368	1.7459	2.1199	2.5835	2.9208
17	1.3334	1.7396	2.1098	2.5669	2.8982
18	1.3304	1.7341	2.1009	2.5524	2.8784
19	1.3277	1.7291	2.0930	2.5395	2.8609
20	1.3253	1.7247	2.0860	2.5280	2.8453
21	1.3232	1.7207	2.0796	2.5176	2.8314
22	1.3212	1.7171	2.0739	2.5083	2.8188
23	1.3195	1.7139	2.0687	2.4999	2.8073
24	1.3178	1.7109	2.0639	2.4922	2.7969
25	1.3163	1.7081	2.0595	2.4851	2.7874
26	1.3150	1.7056	2.0555	2.4786	2.7787
27	1.3137	1.7033	2.0518	2.4727	2.7707
28	1.3125	1.7011	2.0484	2.4671	2.7633
29	1.3114	1.6991	2.0452	2.4620	2.7564
∞	1.2816	1.6449	1.9600	2.3263	2.5758

TABLE A.6
POISSON CUMULATIVE DISTRIBUTION FUNCTION

	μ								
r	0.1	0.2	0.3	0.4	0.5	0.6	0.7	0.8	0.9
0	0.9048	0.8187	0.7408	0.6703	0.6065	0.5488	0.4966	0.4493	0.4066
1	0.9953	0.9825	0.9631	0.9384	0.9098	0.8781	0.8442	0.8088	0.7725
2	0.9998	0.9989	0.9964	0.9921	0.9856	0.9769	0.9659	0.9526	0.9371
3	1.0000	0.9999	0.9997	0.9992	0.9982	0.9966	0.9942	0.9909	0.9865
4	1.0000	1.0000	1.0000	0.9999	0.9998	0.9996	0.9992	0.9986	0.9977
5	1.0000	1.0000	1.0000	1.0000	1.0000	1.0000	0.9999	0.9998	0.9997
6	1.0000	1.0000	1.0000	1.0000	1.0000	1.0000	1.0000	1.0000	1.0000

	μ								
r	1.0	1.5	2.0	2.5	3.0	3.5	4.0	4.5	5.0
0	0.3679	0.2231	0.1353	0.0821	0.0498	0.0302	0.0183	0.0111	0.0067
1	0.7358	0.5578	0.4060	0.2873	0.1991	0.1359	0.0916	0.0611	0.0404
2	0.9197	0.8088	0.6767	0.5438	0.4232	0.3208	0.2381	0.1736	0.1247
3	0.9810	0.9344	0.8571	0.7576	0.6472	0.5366	0.4335	0.3423	0.2650
4	0.9963	0.9814	0.9473	0.8912	0.8153	0.7254	0.6288	0.5321	0.4405
5	0.9994	0.9955	0.9834	0.9580	0.9161	0.8576	0.7851	0.7029	0.6160
6	0.9999	0.9991	0.9955	0.9858	0.9665	0.9347	0.8893	0.8311	0.7622
7	1.0000	0.9998	0.9989	0.9958	0.9881	0.9733	0.9489	0.9134	0.8666
8	1.0000	1.0000	0.9998	0.9989	0.9962	0.9901	0.9786	0.9597	0.9319
9	1.0000	1.0000	1.0000	0.9997	0.9989	0.9967	0.9919	0.9829	0.9682
10	1.0000	1.0000	1.0000	0.9999	0.9997	0.9990	0.9972	0.9933	0.9863

	μ								
r	6	7	8	9	10	12	15	20	25
0	0.0025	0.0009	0.0003	0.0001	0.0000	0.0000	0.0000	0.0000	0.0000
1	0.0174	0.0073	0.0030	0.0012	0.0005	0.0001	0.0000	0.0000	0.0000
2	0.0620	0.0296	0.0138	0.0062	0.0028	0.0005	0.0000	0.0000	0.0000
3	0.1512	0.0818	0.0424	0.0212	0.0103	0.0023	0.0002	0.0000	0.0000
4	0.2851	0.1730	0.0996	0.0550	0.0293	0.0076	0.0009	0.0000	0.0000
5	0.4457	0.3007	0.1912	0.1157	0.0671	0.0203	0.0028	0.0001	0.0000
6	0.6063	0.4497	0.3134	0.2068	0.1301	0.0458	0.0076	0.0003	0.0000
7	0.7440	0.5987	0.4530	0.3239	0.2202	0.0895	0.0180	0.0008	0.0000
8	0.8472	0.7291	0.5925	0.4557	0.3328	0.1550	0.0374	0.0021	0.0001
9	0.9161	0.8305	0.7166	0.5874	0.4579	0.2424	0.0699	0.0050	0.0002
10	0.9574	0.9015	0.8159	0.7060	0.5830	0.3472	0.1185	0.0108	0.0006
11	0.9799	0.9467	0.8881	0.8030	0.6968	0.4616	0.1848	0.0214	0.0014
12	0.9912	0.9730	0.9362	0.8758	0.7916	0.5760	0.2676	0.0390	0.0031
13	0.9964	0.9872	0.9658	0.9261	0.8645	0.6815	0.3632	0.0661	0.0065
14	0.9986	0.9943	0.9827	0.9585	0.9165	0.7720	0.4657	0.1049	0.0124
15	0.9995	0.9976	0.9918	0.9780	0.9513	0.8444	0.5681	0.1565	0.0223
16	0.9998	0.9990	0.9963	0.9889	0.9730	0.8987	0.6641	0.2211	0.0377
17	0.9999	0.9996	0.9984	0.9947	0.9857	0.9370	0.7489	0.2970	0.0605
18	1.0000	0.9999	0.9993	0.9976	0.9928	0.9626	0.8195	0.3814	0.0920
19	1.0000	1.0000	0.9997	0.9989	0.9965	0.9787	0.8752	0.4703	0.1336
20	1.0000	1.0000	0.9999	0.9996	0.9984	0.9884	0.9170	0.5591	0.1855
21	1.0000	1.0000	1.0000	0.9998	0.9993	0.9939	0.9469	0.6437	0.2473
22	1.0000	1.0000	1.0000	0.9999	0.9997	0.9970	0.9673	0.7206	0.3175
23	1.0000	1.0000	1.0000	1.0000	0.9999	0.9985	0.9805	0.7875	0.3939
24	1.0000	1.0000	1.0000	1.0000	1.0000	0.9993	0.9888	0.8432	0.4734

Appendix B

Government Documents

The United States government used to publish government documents. The following documents are available on the accompanying CD:

MIL-HDBK-189	Reliability Growth Management
MIL-HDBK-217	Reliability Prediction of Electronic Equipment
MIL-HDBK-251	Reliability/Design Thermal Applications
MIL-HDBK-263	Electrostatic Discharge Control Handbook for Protection of Electrical and Electronic Parts, Assemblies, and Equipment
MIL-HDBK-338	Electronic Reliability Design Handbook
MIL-HDBK-472	Maintainability Prediction
MIL-HDBK-781	Reliability Test Methods, Plans, and Environments for Engineering Development, Qualification, and Production
MIL-STD-690	Failure Rate Sampling Plans and Procedures
MIL-STD-721	Definition of Terms for Reliability and Maintainability
MIL-STD-785	Reliability Program for Systems and Equipment, Development and Production
MIL-STD-883	Test Methods and Procedures for Microelectronics
MIL-STD-965	Parts Control Program
MIL-STD-1472	Human Engineering Design Criteria for Military Systems, Equipment, and Facilities
MIL-STD-1629	Procedures for Performing a Failure Mode, Effects, and Criticality Analysis
MIL-STD-1635	Reliability Growth Testing
MIL-STD-2074	Failure Classification for Reliability Testing

Other documents of interest include the following:

DoD-HDBK-344 (U.S.A.F.), *Environmental Stress Screening of Electronic Equipment*, U.S. Department of Defense, 1993.

Environmental Stress Screening of Electronic Hardware (ESSEH), Institute of Environmental Sciences and Technology (IEST), 1990.

LC-78-2 Storage Reliability Analysis Report (Vol. 1, Electrical and Electronic Devices; Vol. 2, Electromechanical Devices; Vol. 3, Hydraulic and Pneumatic Devices; Vol. 4, Ordnance Devices; Vol. 5, Optical Devices and Electro-Optical Devices)

MIL-STD-280 Definitions of Item Levels, Item Exchangeability, Models and Related Items, U.S. Department of Defense, 1990.

MIL-STD-470 Maintainability Program Requirements for Systems and Equipment

MIL-STD-756 Reliability Modeling and Prediction

MIL-STD-757 Reliability Evaluation from Demonstration Data

MIL-STD-781 Reliability Testing for Engineering Development, Qualification, and Production

MIL-STD-790 Reliability Assurance for Electronic Parts

MIL-STD-810 Environmental Test Methods and Engineering Guidelines

MIL-STD-2068 Reliability Development Tests

NAVMAT P-9492 Navy Manufacturing Screening Program, U.S. Department of Navy, May 1979.

NPRD-2 Nonelectric Parts Reliability Data

RADC-TR-75-22 Nonelectric Reliability Notebook

Rome Air Development Center, various studies, Griffiss AFB, Rome, NY 13441-5700.

Appendix C

Glossary

Accessibility—A measure of the relative ease of admission to the various areas of an item for the purpose of operation or maintenance.

Achieved—Obtained as the result of measurement.

Alignment—Performing the adjustments that are necessary to return an item to specified operation.

Availability—A measure of the degree to which an item is in an operable and committable state at the start of a mission when the mission is called for at an unknown (random) time. (Item state at the start of mission includes the combined effects of the readiness-related system reliability and maintainability [R&M] parameters but excludes mission time; *see* **Dependability**).

Burn-In (Pre-Conditioning)—The operation of an item under stress to stabilize its characteristics; not to be confused with **Debugging**.

Calibration—A comparison of a measuring device with a known standard; not to be confused with **Alignment** (*see* MIL-C-45662).

Chargeable—Within the responsibility of a given organizational entity (applied to terms such as failures and maintenance time).

Checkout—Tests or observations of an item to determine its condition or status.

Corrective Action—A documented design, process, procedure, or materials change implemented and validated to correct the cause of failure or design deficiency.

Criticality—A relative measure of the consequence of a failure mode and its frequency of occurrences.

Debugging—A process to detect and remedy inadequacies; not to be confused with terms such as **Burn-In**, **Fault Isolation**, or **Screening**.

Degradation—A gradual impairment in ability to perform.

Demonstrated—That which has been measured by the use of objective evidence gathered under specified conditions.

Dependability—A measure of the degree to which an item is operable and capable of performing its required function at any (random) time during a specified mission profile, given item availability at the start of the mission. (Item state during a mission includes the combined effects of the mission-related system reliability and maintainability [R&M] parameters but excludes non-mission time; *see* **Availability**.)

Derating—(a) Using an item in such a way that applied stresses are below rated values, or (b) the lowering of the rating of an item in one stress field.

Direct Maintenance Man-Hours per Maintenance Action (DMMH/MA)—A measure of the maintainability parameter related to item demand for maintenance manpower. The sum of direct maintenance man-hours, divided by the total number of maintenance actions (preventive and corrective) during a stated period of time.

Direct Maintenance Man-Hours per Maintenance Event (DMMH/ME)—A measure of the maintainability parameter related to item demand for maintenance manpower. The sum of direct maintenance man-hours, divided by the total number of maintenance events (preventive and corrective) during a stated period of time.

Disassemble—Opening an item and removing a number of parts or subassemblies to make the item that is to be replaced accessible for removal. This does not include the actual removal of the item to be replaced.

Dormant—The state wherein an item is able to function but is not required to function; not to be confused with downtime operable, which is the state of being able to perform the intended function; the same as **Not Operating**.

Downing Event—The event that causes an item to become unavailable to initiate its mission (the transition from uptime to downtime).

Durability—A measure of useful life (a special case of reliability).

Environment—The aggregate of all external and internal conditions (e.g., temperature, humidity, radiation, magnetic and electric fields, shock vibration), whether natural or man-made, or self-induced, that influences the form, performance, reliability, or survival of an item.

Environmental Stress Screening (ESS)—A series of tests conducted under environmental stresses to disclose weak parts and workmanship defects for correction.

Failure—The event, or inoperable state, in which any item or part of an item does not or would not perform as previously specified.

Failure Analysis—Subsequent to a failure, the logical systematic examination of an item, its construction, application, and documentation to identify the failure mode and to determine the failure mechanism and its basic course.

Failure Catastrophic—A failure that can cause item loss.

Failure Critical—A failure, or combination of failures, that prevents an item from performing a specified mission.

Failure Dependent—Failure that is caused by the failure of an associated item(s); not **Failure Independent**.

Failure Effect—The consequence(s) a failure mode has on the operation, function, or status of an item. Failure effects are classified as local effect, next higher level, and end effect.

Failure, Independent—Failure that occurs without being caused by the failure of any other item; not **Failure Dependent**.

Failure, Intermittent—Failure for a limited period of time, followed by the item's recovery of its ability to perform within specified limits without any remedial action.

Failure Mechanism—The physical, chemical, electrical, thermal, or other process that results in failure.

Failure Mode—The consequence of the mechanism through which the failure occurs (i.e., short, open, fracture, excessive wear).

Failure Mode and Effects Analysis (FMEA)—A procedure by which each potential failure mode in a system is analyzed to determine the results or effects thereof on the system, and to classify each potential failure mode according to its severity.

Failure, Non-Chargeable—(a) A non-relevant failure, or (b) a relevant failure caused by a condition previously specified as not within the responsibility of a given organizational entity. (All relevant failures are chargeable to one organizational entity or another.)

Failure, Non-Relevant—(a) A failure verified as having been caused by a condition not present in the operational environment, or (b) a failure verified as peculiar to an item design that will not enter the operational inventory.

Failure, Random—Failure whose occurrence is predictable only in a probabilistic or statistical sense. This applies to all distributions.

Failure Rate—The total number of failures within an item population, divided by the total number of life units expended by that population, during a particular measurement interval under stated conditions.

Fault—Immediate cause of failure (e.g., maladjustment, misalignment, defect).

Fault Isolation—The process of determining the location of a fault to the extent necessary to effect repair.

Fault Localization—The process of determining the approximate location of a fault.

Inherent R&M Value—A measure of reliability or maintainability (R&M) that includes only the effects of an item design and its application, and assumes an ideal operation and support environment.

Interchange—Removing the item that is to be replaced, and installing the replacement item.

Inventory, Active—The group of items assigned to an operational status.

Inventory, Inactive—The group of items being held in reserve for possible future assignments to an operational status.

Item—A non-specific term used to denote any product, including systems, materials parts, subassemblies, sets, and accessories. (Source: MIL-STD-280.)

Life Profile—A time-phased description of the events and environments an item experiences from manufacture to final expenditures of removal from the operational inventory, to include one or more mission profiles.

Life Units—A measure of use duration applicable to the item (e.g., operating hours, cycles, distance, rounds fired, attempts to operate).

Maintainability—The measure of the ability of an item to be retained in or restored to specified condition when maintenance is performed by personnel having specified skill levels, using prescribed procedures and resources, at each prescribed level of maintenance and repair.

Maintainability, Mission—The measure of the ability of an item to be retained in or restored to specified condition when maintenance is performed during the course of a specified mission profile (the mission-related system maintainability parameter).

Maintenance—All actions necessary for retaining an item in or restoring it to a specified condition.

Maintenance Action—An element of a maintenance event. One or more tasks (i.e., fault localization, fault isolation, servicing, inspection) necessary to retain an item in or restore it to a specified condition.

Maintenance, Corrective—All actions performed as a result of failure, to restore an item to a specified condition. Corrective maintenance can include any or all of the following steps: localization, isolation, disassembly, interchange, reassembly, alignment, and checkout.

Maintenance Event—One or more maintenance actions required to effect corrective and preventive maintenance due to any type of failure or malfunction, false alarm, or scheduled maintenance plan.

Maintenance Manning Level—The total authorized or assigned personnel per system at specified levels of maintenance organization.

Maintenance, Preventive—All actions performed in an attempt to retain an item in specified condition by providing systematic inspection, detection, and prevention of incipient failures.

Maintenance Ratio—A measure of the total maintenance manpower burden required to maintain an item. It is expressed as the cumulative number of man-hours of maintenance expended in direct labor during a given period of the life units divided by the cumulative number of end-item life units during the same period.

Maintenance, Scheduled—Preventive maintenance performed at prescribed points during the life of an item.

Maintenance Time—An element of downtime that excludes modification and delay time.

Maintenance, Unscheduled—Corrective maintenance required by item conditions.

Malfunction—The event, or inoperable state, in which any item or part of an item does not, or would not, perform as previously specified; the same as **Failure**.

Mean Maintenance Time—The measure of item maintainability, taking into account maintenance policy. The sum of preventive and corrective maintenance times, divided by the sum of scheduled and unscheduled maintenance events, during a stated period of time.

Mean Time Between Demands (MTBD)—A measure of the system reliability parameter related to demand for logistic support. The total number of system life units divided by the total number of item demands on the supply system during a stated period of time, such as shop replaceable unit (SRU), weapon replaceable unit (WRU), line replacement unit (LRU), and shop replaceable assembly (SRA).

Mean Time Between Downing Events (MTBDE)—A measure of the system reliability parameter related to availability and readiness. The total number of system life units, divided by the total number of events in which the system becomes unavailable to initiate its mission(s), during a stated period of time.

Mean Time Between Failures (MTBF)—A basic measure of reliability for repairable items. The mean number of life units during which all parts of the item perform within their specified limits, during a particular measurement interval under stated conditions.

Mean Time Between Maintenance (MTBM)—A measure of reliability, taking into account maintenance policy. The total number of life units expended by a given time, divided by the total number of maintenance events (scheduled and unscheduled) due to that item.

Mean Time Between Maintenance Actions (MTBMA)—A measure of the system reliability parameter related to demand for maintenance manpower. The total number of system life units, divided by the total number of maintenance actions (preventive and corrective) during a stated period of time.

Mean Time Between Removals (MTBR)—A measure of the system reliability parameter related to demand for logistic support. The total number of system life units divided by the total number of items removed from that system during a stated period of time. This term is defined to exclude removals performed to facilitate other maintenance and removals for product improvement.

Mean Time to Failure (MTTF)—A basic measure of reliability for non-repairable items. The total number of life units of an item divided by the total number of failures within that population, during a particular measurement interval under stated conditions.

Mean Time to Repair (MTTR)—A basic measure of maintainability. The sum of corrective maintenance times at any specific level of repair, divided by the total number of failures within an item repaired at that level, during a particular interval under stated conditions.

Mean Time to Restore System (MTTRS)—A measure of the system maintainability parameter related to availability and readiness. The total corrective maintenance time, associated with downing events, divided by the total number of downing events, during a stated period of time. Excludes time for off-system maintenance and repair of detached components.

Mean Time to Service (MTTS)—A measure of an on-system maintainability characteristic related to servicing that is calculated by dividing the total scheduled crew, operator, and driver servicing time by the number of times the item was serviced.

Mission Profile—A time-phased description of the events and environments an item experiences from initiation to completion of a specified mission, to include the criteria of mission success or critical failures.

Mission Time Between Critical Failures (MTBCF)—A measure of mission reliability. The total amount of mission time, divided by the total number of critical failures during a stated series of missions.

Mission Time to Restore Functions (MTTRF)—A measure of mission maintainability. The total corrective critical failure maintenance time, divided by the total number of critical failures, during the course of a specified mission profile.

Not Operating (Dormant)—The state wherein an item is able to function but is not required to function; not to be confused with **Time, Down (Downtime)**.

Operable—The state of being able to perform the intended function.

Operational Readiness—The ability of a military unit to respond to its operation plan(s) upon receipt of an operations order (function of assigned strength, item availability, status, supply, training, etc.).

Operational R&M Value—A measure of reliability or maintainability (R&M) that includes the combined effects of item design, installation, quality, environment, operation, maintenance, and repair.

Predicted—That which is expected at some future time, postulated on analysis of past experience and tests.

Reassembly—Assembling the items that were removed during disassembly and closing the reassembled items.

Redundancy—The existence of more than one means for accomplishing a given function. Each means of accomplishing the function need not necessarily be identical.

Redundancy, Active—That redundancy wherein all redundant items are operating simultaneously.

Redundancy, Standby—That redundancy wherein the alternative means of performing the function is not operating until it is activated upon failure of the primary means of performing the function.

Reliability—(a) The duration or probability of failure-free performance under stated conditions, or (b) the probability that an item can perform its intended function for a specified interval under stated conditions. (For non-redundant items, this is equivalent to definition (a). For redundant items, this is equivalent to the definition of mission reliability.)

Reliability Growth—The improvement in a reliability parameter caused by the successful correction of deficiencies in item design or manufacture.

Reliability Mission—The ability of an item to perform its required functions for the duration of a specified mission profile.

Repairable Item—An item that can be restored by corrective maintenance to perform all of its required functions.

R&M Accounting—That set of mathematical tasks that establish and allocate quantitative reliability and maintainability (R&M) requirements, and predict and measure quantitative R&M requirements.

R&M Engineering—That set of design, development, and manufacturing tasks by which reliability and maintainability (R&M) are achieved.

Screening—A process for inspecting items to remove those that are unsatisfactory or those likely to exhibit early failure. Inspection includes visual examination, physical dimension measurement, and functional performance measurement under specified environmental conditions.

Servicing—The performance of any act needed to keep an item in operating condition (i.e., lubricating, fueling, oiling, cleaning), but not including preventive maintenance of parts or corrective maintenance tasks.

Single-Point Failure—The failure of an item that would result in failure of the system and is not compensated for by redundancy or alternative operational procedure.

Sneak Circuit Analysis—A procedure conducted to identify latent paths that cause occurrence of unwanted functions or inhibit desired functions, assuming all components are functioning properly.

Storage Life (Shelf Life)—The length of time an item can be stored under specified conditions and still meet specified requirements.

Subsystem—A combination of sets, groups, and so forth that performs an operational function within a system and is a major subdivision of the system (e.g., data processing subsystem, guidance subsystem). (Source: MIL-STD-280.)

System, General—A composite of equipment and skills, and techniques capable of performing or supporting an operational role, or both. A complete system includes all equipment, related facilities, material, software, services, and personnel required for its operation and support to the degree that it can be considered self-sufficient in its intended operational environment.

System R&M Parameter—A measure of reliability and maintainability (R&M) in which the units of measurement are directly related to operational readiness, mission success, maintenance manpower cost, or logistic support cost.

Test, Acceptance—A test conducted under specified conditions by, or on behalf of, the government, using delivered or deliverable items, to determine the compliance of the item with specified requirements (including acceptance of first production units).

Test Measurement and Diagnostic Equipment (TMDE)—Any system or device used to evaluate the condition of an item to identify or isolate any actual or potential failures.

Test, Qualification (Design Approval)—A test conducted under specified conditions, by or on behalf of the government, using items representative of the production configuration, in order to determine compliance with item design requirements as a basis for production approval; also known as demonstration.

Testing Development (Growth)—A series of tests conducted to disclose deficiencies and to verify that corrective actions will prevent recurrence in the operational inventory. Note that repair of test items does not constitute correction of deficiencies; also known as test-analyze-and-fix (TAAF) testing.

Time—The universal measure of duration. The general word "time" will be modified by an additional term when used in reference to operating time, mission time, test time, and so forth. In general expressions such as "mean time between failure (MTBF)," time stands for "life units," which must be more specifically defined whenever the general term refers to a particular item.

See Figure C.1 for time relationships.

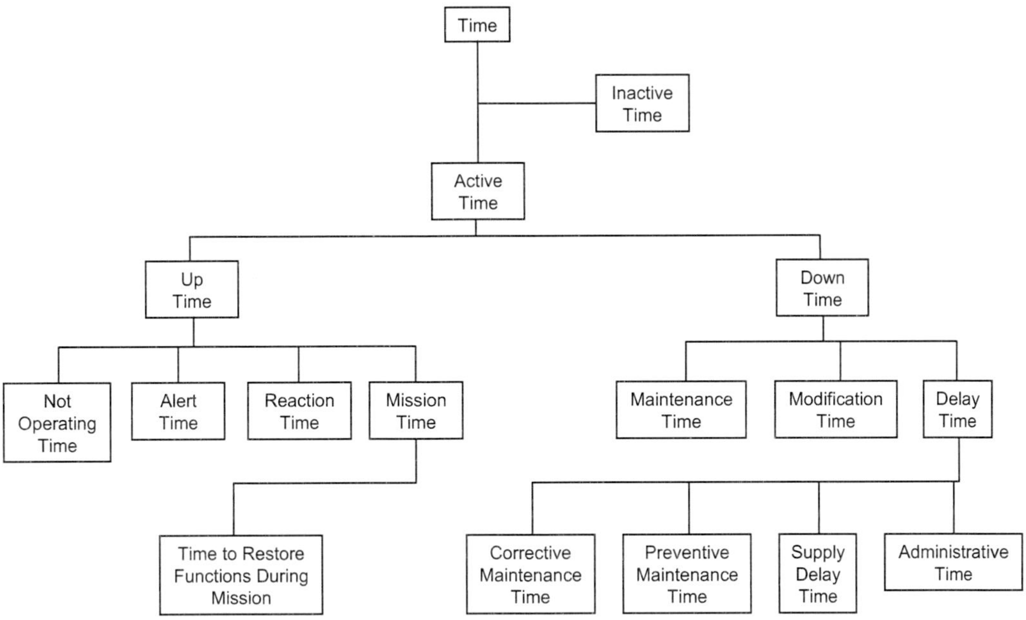

Figure C.1 *Time relationships.*

Time, Active—That time during which an item is in an operational inventory.

Time, Administrative—That element of delay time, not included in the supply.

Time, Alert—That element of uptime during which an item is assumed to be in specified operating condition and is awaiting a command to perform its intended mission.

Time, Checkout—That element of maintenance time during which performance of an item is verified to be a specified condition.

Time, Delay—That element of downtime during which no maintenance is being accomplished on the item because of either supply or administrative delay.

Time, Down (Downtime)—That element of active time during which an item is not in condition to perform its required function (reduces **Availability** and **Dependability**).

Time, Inactive—That time during which an item is in reserve.

Time, Mission—That element of uptime required to perform a stated mission profile.

Time, Modification—The time necessary to introduce any specific change(s) to an item to improve its characteristics or to add new ones.

Time, Not Operating—That element of uptime during which an item is not required to operate.

Time, Reaction—That element of uptime needed to initiate a mission, measured from the time that the command is received.

Time, Supply Delay—That element of delay time during which a needed replacement item is being obtained.

Time, Turnaround—That element of maintenance time needed to replenish consumables and check out an item for recommitment.

Time, Up (Uptime)—That element of active time during which an item is in condition to perform its required functions.

Uptime Ratio—A composite measure of operational availability and dependability that includes the combined effects of item design, installation, quality, environment, operation, maintenance, repair, and logistic support; the quotient of uptime divided by uptime plus downtime.

Useful Life—The number of life units from manufacture to when the item has an unrepairable failure or unacceptable failure rate.

Utilization Rate—The planned or actual number of life units expended, or missions attempted, during a stated interval of calendar time.

Wearout—The process that results in an increase of the failure rate or probability of failure with increasing number of life units.

Appendix D

List of Acronyms

The following acronyms are used throughout this book:

AMSAA	Army Material Systems Analysis Activity
DMMH/MA	Direct maintenance man-hours per maintenance action
DMMH/ME	Direct maintenance man-hours per maintenance event
EMI	Electromagnetic interference
ESD	Electrostatic discharge
ESS	Environmental stress screening
ESSEH	*Environmental Stress Screening of Electronic Hardware*
FMEA	Failure modes and effects analysis
FRACAS	Failure rate analysis and corrective action system
GPA	Grade point average
GRMS	g-force root mean square
HALT	Highly accelerated life testing
HASS	Highly accelerated stress screen
HP	Hewlett-Packard
IC	Integrated circuit
IES	Institute of Environmental Sciences

I/O	Input and output
LRU	Line replacement unit
MTBCF	Mission time between critical failures
MTBD	Mean time between demands
MTBDE	Mean time between downing events
MTBF	Mean time between failures
MTBM	Mean time between maintenance
MTBMA	Mean time between maintenance actions
MTBR	Mean time between removals
MTTF	Mean time to failure
MTTR	Mean time to repair
MTTRF	Mission time to restore functions
MTTRS	Mean time to restore system
MTTS	Mean time to service
OC	Operating characteristic (curve)
PSD	Power spectral density
RADC	Rome Air Development Center
RIW	Reliability improvement warranty (contract)
R&M	Reliability and maintainability
ROI	Return on investment
SPC	Statistical process control (program)
SRA	Shop replaceable assembly
SRU	Shop replaceable unit

TAAF	Test-analyze-and-fix (testing)
TMDE	Test measurement and diagnostic equipment
WRU	Weapon replaceable unit

REFERENCES

Accelerated Testing Research: Special Publications (2000), Society of Automotive Engineers, Warrendale, PA.

Chan, H. Anthony (2001). *Accelerated Stress Testing Handbook: Guide for Achieving Quality Products*, IEEE Press, Institute of Electrical and Electronics Engineers, Piscataway, NJ.

Dhillon, B.S. (1985). *Quality Control, Reliability, and Engineering Design*, Marcel Dekker, New York.

Dodson, B.L., and M.D. Mulcahy (1992). *Certified Reliability Engineer Examination Study Guide*, Quality Publishing, Tucson, AZ.

Dodson, Bryan (1994). *Weibull Analysis (with Software)*, ASQC Quality Press, Milwaukee, WI.

Dovich, R.A. (1990). *Reliability Statistics*, ASQC Quality Press, Milwaukee, WI.

Environmental Stress Screening Handbook (2004), Thermotron Ltd., Holland, MI.

Fundamentals of Accelerated Stress Testing (2004), Thermotron Ltd., Holland, MI.

Hobbs, Gregg K. (2000). *Accelerated Reliability Engineering: HALT and HASS*, John Wiley & Sons, New York.

Ireson, G.W., and C.F. Coombs (1996). *Handbook of Reliability Engineering and Management*, McGraw-Hill, New York.

Kapur, K.C., and L.R. Lamberson (1977). *Reliability in Engineering Design,* John Wiley & Sons, New York.

Kielpinski, T.J., and W. Nelson (1975). "Optimum Censored Accelerated Life Tests for Normal and Lognormal Life Distributions," *IEEE Transactions on Reliability*, R-24:310–320, Institute of Electrical and Electronics Engineers, Piscataway, NJ.

Krishnamoorthi, K.S. (1992). *Reliability Methods for Engineers*, ASQ Quality Press, Milwaukee, WI.

Lall, Pradeep, Michael G. Pecht, and Edward B. Hakim, (1997). *Influence of Temperature on Microelectronics*, CRC Press, Boca Raton, FL.

Lewis, E.E. (1995). *Introduction to Reliability Engineering*, John Wiley & Sons, New York.

Meeker, W., and W. Nelson (1976). "Optimum Accelerated Life Tests for Weibull and Extreme Value Distributions," *IEEE Transactions on Reliability*, R-25:20–24, Institute of Electrical and Electronics Engineers, Piscataway, NJ.

Meeker, W.Q., and G.J. Hahn (1985). *How to Plan an Accelerated Life Test—Some Practical Guidelines*, ASQ Quality Press, Milwaukee, WI.

Nelson, W. (1990). *Accelerated Testing: Statistical Models, Test Plans and Data Analysis*, John Wiley & Sons, New York.

NIST/SEMATECH e-Handbook of Statistical Methods, http://www.itl.nist.gov/div898/handbook/.

O'Connor, Patrick D.T. (2002). *Practical Reliability Engineering*, John Wiley & Sons, New York.

Shooman, M.L. (1990). *Probabilistic Reliability: An Engineering Approach,* Robert E. Krieger, Malabar, FL.

Staudte, Robert G., and S.J. Sheather (1990). *Robust Estimation and Testing*, John Wiley & Sons, New York.

Tobias, P.A., and D.C. Trindade (1995). *Applied Reliability*, Van Nostrand Reinhold, New York.

U.S. Army Materiel Systems Analysis Activity (1999). *Reliability Growth Handbook.*

U.S. Department of Defense. MIL-HDBK-189: *Reliability Growth Management*, Naval Publications and Forms Center, Philadelphia.

U.S. Department of Defense. MIL-HDBK-781: *Reliability Test Methods, Plans and Environments for Engineering Development, Qualification and Production*, Naval Publications and Forms Center, Philadelphia.

U.S. Department of Defense. MIL-STD-810: *Environmental Test Methods and Engineering Guidelines*, Naval Publications and Forms Center, Philadelphia.

U.S. Department of Defense. MIL-STD-1635: *Reliability Growth Testing*, Naval Publications and Forms Center, Philadelphia.

Walker, N. Edward (1998). *The Design Analysis Handbook*, Newnes, Boston.

Young, W.R. (1998). "Accelerated Temperature Pharmaceutical Product Stability Determinations," *Drug Development and Industrial Pharmacy*.

INDEX

About the Authors

Bryan Dodson is the Director of Reliability Engineering at Visteon in Dearborn, Michigan. He formerly was the Manager of Reliability Engineering and Quality Information Systems for Continental Teves, Brake and Chassis, North America. Prior to joining Continental Teves, Dr. Dodson held the positions of Total Quality Management (TQM) Leader and Reliability Leader at Alcoa and the position of Industrial Engineer at Morton Thiokol.

Dr. Dodson has authored seven books and five commercial software programs, including the *Reliability Engineering Handbook*, *Weibull Analysis: with Software*, and *Certified Reliability Engineer Examination Study Guide*. He also is the author of the International Quality Federation's Six Sigma certification exam.

Dr. Dodson holds a B.S. in petroleum engineering, an M.S. in industrial engineering, an M.B.A., and a Doctor of Business Administration. In addition, he is a Certified Quality Engineer (CQE), a Certified Reliability Engineer (CRE), a Six Sigma Master Black Belt, and a licensed Professional Engineer in Michigan and Indiana.

Harry Schwab has almost forty years of experience as an engineer, specializing in structural dynamics and vibrations. He holds a Bachelor of Aerospace Engineering from the Georgia Institute of Technology, an M.S. in Engineering Mechanics from the University of Missouri—Rolla, and an M.B.A. from the Florida Institute of Technology.

Mr. Schwab has worked in the aerospace, automotive, and commercial industries. The products with which he has been involved include automobiles, trucks, railroad freight cars, aircraft, boats, and guided missiles. He has many years of experience in working with random vibrations and developed a method for deriving laboratory testing environments from non-stationary data. This technique can be used to accurately simulate real-world environments while greatly decreasing test time.

Mr. Schwab has authored several technical papers for SAE International and the Institute of Environmental Sciences and Technology (IEST), as well as for other technical organizations and publications. These cover the fields of vibration testing and analysis. He developed innovative testing techniques for both the aerospace and automotive industries, and he is a registered Professional Engineer in Missouri and Florida.

W ERRATA

Accelerated Testing

A Practitioner's Guide to Accelerated and Reliability Testing (R-304)

Corrections to equations and figures:

Chapter 5

$$b = \frac{\beta \ln\left(\frac{\theta_0}{\theta_1}\right)}{a} \tag{5.23}$$

3rd equation:

$$b = \frac{(2.2)\ln\left(\frac{50}{35}\right)}{0.00021799} = 13{,}259$$

Figure 5.17

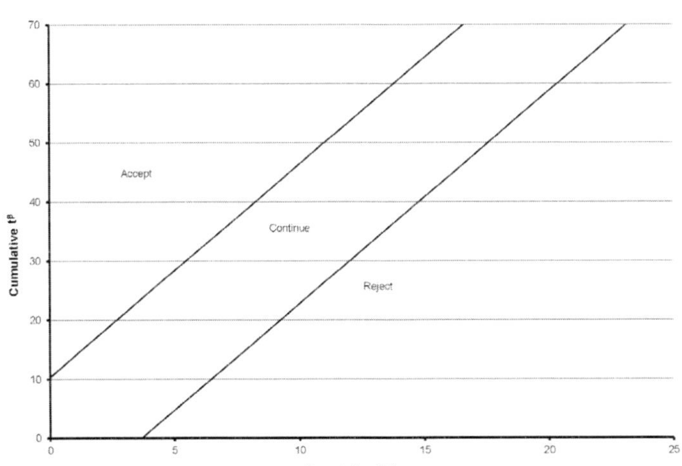